101 Optoelectronic Projects

No. 3205
$24.95

101 Optoelectronic Projects

Delton T. Horn

TAB BOOKS Inc.
Blue Ridge Summit, PA

FIRST EDITION
FIRST PRINTING

Copyright © 1989 by TAB BOOKS Inc.
Printed in the United States of America

Library of Congress Cataloging in Publication Data

Horn, Delton T.
    101 optoelectronic projects / by Delton T. Horn.
      p. cm.
    Includes index.
    ISBN 0-8306-9205-3   ISBN 0-8306-3205-0 (pbk.)
    1. Optoelectronics—Amateurs' manuals. I. Title. II. Title: One
  hundred one optoelectronic projects. III. Title: One hundred and
  one optoelectronic projects.
  TA1750.H67 1989
  621.381—dc19   89-4273
                                                     CIP

TAB BOOKS Inc. offers software for sale. For information and a catalog, please contact TAB Software Department, Blue Ridge Summit, PA 17294-0850.

Questions regarding the content of this book should be addressed to:

    Reader Inquiry Branch
    TAB BOOKS Inc.
    Blue Ridge Summit, PA 17294-0214

Roland Phelps: Acquisitions Editor
Steven L. Burwen: Technical Editor
Katherine Brown: Production
Lori E. Schlosser: Cover Design

Contents

11 Communications Projects   155

12 Photography and Light-Meter Projects   171

13 Counter Circuits   185

14 Miscellaneous Projects   207

In Conclusion 224

Index 225

_____Introduction_____

No electronic circuit is going to be of much use to anybody unless it has some way of exchanging signals or information with the outside world. One way to get signals into or out of a circuit is to use light. Devices that can use light either as an input or an output are known as *optoelectric devices*.

Optoelectric devices that are used for output signals produce light in response to specific signal conditions within the circuit. For input purposes, a *photosensor* is used. A photosensor is an optoelectric device that varies some electrical parameter in response to the light shining on it.

At first glance, optoelectric devices might seem rather trivial, with only limited practical applications, but this is not the case at all. In fact, optoelectrics have countless potential applications. This book features over 100 projects using optoelectric components in a variety of ways for many different purposes. You are bound to find several circuits you can use in this book.

Part I

COMPONENTS

Chapter 1
Photocells
and
Photoresistors

Optoelectric devices can be divided into two very broad categories: they either are light sensors, or light generators. The prefix "opto-" is short for *optical*, so *optoelectric* means both light and electricity are involved.

I will begin with a little theory and a general examination of basic optoelectric devices. In the first two chapters, we will look at light sensor devices. Chapter 3 covers light generator devices. Optoisolators incorporate both a light generator and a light sensor in a single package. I cover these devices in Chapter 4. The actual projects begin in Chapter 5.

This chapter covers the two most basic types of light sensors: photocells and photoresistors. Such devices are commonly said to be photosensitive. The prefix "photo-" means *light*, so this term just says that these devices are sensitive to light.

In some texts, photoresistors are sometimes called *photocells*, but this can create confusion. In this book, the term *photocell* is always used to mean a photovoltaic cell.

PHOTOSENSITIVITY IN SEMICONDUCTORS

Both photocells and photoresistors are two-terminal semiconductor devices.

Almost all semiconductor materials are photosensitive to some degree. If a semiconductor is exposed to light, the intensity of that light affects how much current flows through the semiconductor.

For most applications, this photosensitivity would be highly undesirable because lighting is usually an uncontrolled variable, at least as far as most electronic equipment is concerned. To protect against unwanted photosensitive reactions, transistors and integrated circuits are generally enclosed in light-tight metal or black plastic housings. Therefore, when using these components, we can totally ignore the photosensitivity of the semiconductor materials making up the active portion of the devices.

However, in some special applications, we can make good use of the natural photosensitive property of semiconductors. This is where optoelectric sensors come in.

The Photoelectric Effect

All optoelectric sensors utilize the *photoelectric effect*, which is really just another name for the photosensitivity of the semiconductor material (silicon). When the material is exposed to light, it emits electrons.

Actually, all substances exhibit the photoelectric effect to some degree, but for most materials, the number of emitted electrons is negligible. For all practical purposes, a photoelectric material is one that emits significant numbers of electrons when it is exposed to a light source.

There are several variations on the basic photoelectric effect. Different photosensitive materials are best suited for applications within one of these variants.

The basic photoelectric effect, in which a current is emitted in response to light, can also be called *photoemission*. Alkali metals such as cesium, sodium, and potassium are good examples of photoemissive materials.

Photovoltaic cells utilize a related phenomenon known as the *photovoltaic effect*. The name is pretty suggestive. When a photovoltaic material is exposed to light, it generates a voltage. Most photovoltaic cells are made of selenium or silicon.

A similar effect is *photoconductivity*. A photoconductive material decreases its resistance as the light intensity increases. Cadmium sulfide is a popular material for making photoconductive devices. I discuss photoconductivity more thoroughly when we get to the section on photoresistors.

PHOTOVOLTAIC CELLS

One common type of simple optoelectric sensor is the photovoltaic cell, or photocell. This device is also widely known as a solar battery, especially when multiple units are used together. (To be precise, a battery is made up of multiple cells. A common D or AA battery isn't really a battery at all, but just a cell.) Some sources call photovoltaic cells *self-generating photocells*.

The schematic symbol for a photovoltaic cell is shown in Fig. 1-1. Note that this symbol is very similar to the one used to indicate ordinary dry cells (or "batteries").

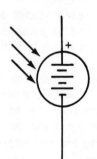

Fig. 1-1. A photovoltaic cell produces a voltage when it is illuminated.

This similarity implies that photocells are used for the same sort of functions. A photocell behaves in a circuit as a dc voltage source.

A photocell is basically just an optically exposed pn junction. Usually silicon is used in the manufacture of this type of device. Silicon is commonly used out of all the photosensitive materials because it gives a relatively higher output voltage for a given level of illumination.

A photocell is virtually identical in construction to a standard diode. The major difference is that in an ordinary diode the pn junction is shielded, while in the photocell the junction is intentionally exposed to external light sources.

To maximize the photoelectric effect, the pn junction in a photocell is usually spread out into a relatively large, thin plate. This provides the maximum possible contact area, or amount of exposure.

When this silicon surface is shielded from light, no current will flow through the cell. However, when it is exposed to a bright light, a small voltage is generated as a result of the photoelectric effect.

If a photovoltaic cell is hooked up to a load and exposed to light, as illustrated in Fig. 1-2, a current will flow through the photocell and the load circuit. The amount of current that will flow depends on the intensity of the light striking the surface of the photocell. The brighter the light is, the higher the amount of current the photocell can supply to the load.

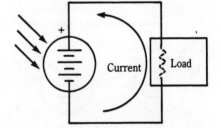

Fig. 1-2. If a photovoltaic cell is hooked up to a load and exposed to light, a current will flow.

The photocell's output voltage, on the other hand, is relatively independent of the light intensity. The voltage provided by most commercially available photovoltaic cells is in the area of one-half volt. Higher voltages can be obtained by connecting multiple photocells in series to create a true battery, as shown in Fig. 1-3.

If the load circuit you want to power from photocells requires a higher current than a single cell can provide, you can add multiple photocells to form a parallel battery. (See Fig. 1-4.)

Whether the photocells are connected in series or in parallel, the combination can be called a *solar battery*. This name is generally used even when artificial lighting is used to activate the photocells.

There is one more important factor that must be kept in mind. The more cells there are in a solar battery, the larger the total surface area must be, and the harder it will be to arrange the cells so that they will all be illuminated more or less evenly. This suggests that generally solar batteries are best suited for fairly low-power circuits.

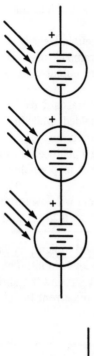

Fig. 1-3. Higher voltages can be obtained by connecting several photovoltaic cells in series.

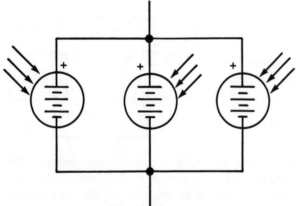

Fig. 1-4. Higher currents can be obtained by connecting several photovoltaic cells in parallel.

High-power photocells have been developed to use solar power as an alternate source of energy. These cells work in a manner similar to the much smaller cells employed in electronics. Specialized materials such as copper oxides are used in these large-scale photocells. To achieve the necessary power levels, the exposed surface area must be much larger per cell. Again, the power output of the solar battery is limited by the space available to mount additional photocells efficiently. This is why solar power has not proven to be very practical for industrial applications, and is not likely to become our primary power source until some major technical breakthroughs are made.

In most applications, a photocell is used as if it was an ordinary battery or dry cell. In all cases you should remember that photovoltaic cells, like any dc voltage source, have a definite *polarity*. That is, one lead is always positive, and the other is always negative. The two leads should never be reversed.

PHOTORESISTORS

Most substances are photoconductive, at least to someextent. Again, in most materials the effect is negligible, but certain semiconductors are more responsive than others.

When a photoconductive substance is illuminated, the charge-carrier mobility is affected. Usually the charge-carrier mobility increases with increasing light levels. The result is that current can flow more easily when a voltage is applied across the photoconductive device. In other words, the resistance decreases.

Photoconductive semiconductors include silicon, germanium, and the sulfides of certain elements. Cadmium sulfide and cadmium selenide are the most commonly used substances for photoconductive applications today.

A photoresistor is a photoconductive component. The conductivity of the semiconductor material (usually cadmium sulfide), varies with the intensity of the light striking it.

The reciprocal of conductivity is resistance; that is,

$$1/U = R$$

or

$$1/R = U$$

where:

U is the conductivity
R is the resistance

From this relationship it is clear that a photoconductive device behaves like a variable resistor, with the light intensity determining the current resistance value. For obvious reasons, photoresistors are often known as *light-dependent resistors,* or LDRs.

The schematic symbol for a photoresistor is shown in Fig. 1-5. Note the similarity to the symbol for an ordinary resistor. The small arrows pointing in towards the resistor indicate light striking the device.

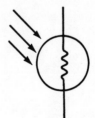

Fig. 1-5. A photoresistor changes its conductivity in response to the illumination level.

Photoresistors generally have two leads, and can usually be employed in place of any ordinary fixed resistor.

Photoresistors are *junctionless* devices. There is no pn junction. The active portion of this component is a continuous piece of semiconductor. Because there is no junction, there is no fixed polarity. Like standard fixed resistors, photoresistors may be installed in a circuit in either direction without affecting circuit operation in any way. You can't get it in backwards.

Most commercially available photoresistors are made from cadmium sulfide and feature a negative resistance/light response. That is, the resistance decreases, as the light increases. When all light is completely blocked from the sensor, typical units exhibit maximum resistances from 1.6 kilohms up to 1000 megohms, depending on the exact design of the individual device. The resistance will decrease as light is shined upon the semiconductor sensor. At 2 footcandles (fc) the resistance will typically be in the range of 1.5K to 700K. If the light intensity is increased to 100 footcandles (fc), the resistance may be as low as 0.11 ohm. This is about the maximum resistance change. Increasing the light intensity beyond this point will not cause the resistance value to drop any further. The exact maximum effective light level (or the upper end of the sensitivity range) will vary with the specific design of the device being used.

Some photoresistors used cadmium selenide instead of cadmium sulfide. These devices generally have a higher resistance for a given level of illumination. Typical dark (completely shielded) resistance values will be from 120K to 3000 MΩ. At 2 fc, the resistance drops from 1.5K to 135K.

A few photoresistors are designed for a positive resistance/light response. That is, the resistance increases with any increase in the illumination. Such devices are fairly rare. Generally, unless otherwise specified, a negative resistance/light response device can be assumed.

In most applications, photoresistors are used as light-controlled potentiometers. Virtually any resistance-determined parameter in almost any circuit can be made light controllable simply by replacing the ordinary resistor with a photoresistor.

Some texts refer to photoresistors as photocells. While, strictly speaking, this is correct, it can cause confusion with photovoltaic cells, so such terminology will be avoided in this book.

_____Chapter 2_____
Phototransistors
and
Related Devices

A number of light-sensitive devices are variations on basic semiconductor components. For example, there are photodiodes, phototransistors, and light-activated silicon-controlled rectifiers (LASCRs). These light-controlled devices operate similarly to their ordinary counterparts, but the light intensity striking the semiconductor material has a direct effect on the component's performance.

PHOTODIODE

The schematic symbol for a photodiode is shown in Fig. 2-1. Note that it is basically similar to the symbol for an ordinary diode, except for the addition of the incoming arrows that symbolize the light striking the sensor.

Fig. 2-1. Schematic symbol for a photodiode.

In some respects, a photodiode is a lot like a photoresistor (see Chapter 1). The effective resistance across the photodiode is controlled by the intensity of the light striking the pn junction. The big difference between a photodiode and a photoresistor is that a photodiode is polarized (because of the pn junction), while a photoresistor is not. In most applications, a photodiode is connected into a circuit so that it is reverse biased.

If it were not for the shielding provided by the component housing, any semiconductor diode (or any pn junction) would function as a photodiode. To prevent unwanted light sensitivity, standard diodes are encased in a light-tight housing. A photodiode, however, is housed in a transparent case. Generally, the pn junction will be spread out over a fairly large area to maximize the exposure to an external light source.

When a photodiode is reverse-biased, the resistance will usually decrease as the light intensity is increased. (A few specialized devices work in the opposite manner.) At some point, the reverse resistance will reach a minimum level, and will not drop any further with additional increases of light intensity. This point is known as the *saturation value* of the device.

In most circuits, photodiodes must be protected against drawing excess current. This can usually be accomplished with a simple current-limiting resistor. An ordinary fixed resistor is used for this purpose.

PHOTOTRANSISTOR

A phototransistor is nothing more than a bipolar transistor with the base/collector junction exposed to external light. Once again, any bipolar transistor exhibits photosensitivity when not enclosed in a light-tight housing.

'Generally, no connection is made to the base of a phototransistor. The intensity of the light striking the sensor serves as the base signal. Most phototransistors do not even have a base lead. These devices have just two leads—the emitter and the collector.

The schematic symbol for a phototransistor is shown in Fig. 2-2. Note that an npn device is shown here. Most phototransistors are of the npn type, although a few pnp units are available.

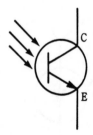

Fig. 2-2. Most phototransistors are of the npn type.

When there is no base lead (or the base lead is not electrically connected), a phototransistor can be biased by placing an LED near the sensor. In the vast majority of applications, such biasing is not necessary. In many cases, it might even be detrimental to the phototransistor's performance in the circuit.

As in the photodiode, resistance in a phototransistor usually decreases with increasing the light intensity, up to the saturation point.

Almost any application that uses an ordinary bipolar transistor can be made light sensitive by substituting a phototransistor. Most phototransistor applications are amplifiers in which the gain is determined by light intensity.

PHOTOFET

Not surprisingly, a photoFET is an FET (field-effect transistor) with an exposed junction (between the gate and the channel) to permit photosensitivity.

Often the gate lead is omitted. The external light intensity is used as the gate signal.

PhotoFETs are quite similar to bipolar phototransistors. The biggest difference is that the photoFET features a higher impedance.

LASCR

A LASCR is a light-activated silicon-controlled rectifier. The name is pretty self-explanatory. This device is used for the same kind of applications that use ordinary SCRs. With a LASCR, an external light source provides the gate current, rather than an electrical source. The schematic symbol for a LASCR is shown in Fig. 2-3.

Fig. 2-3. A LASCR is a light-activated silicon-controlled rectifier.

Normally, the LASCR does not conduct. When the incoming light intensity exceeds a specific level, the LASCR is triggered and starts to conduct. This operation is the same as with an ordinary SCR, which does not conduct until the applied signal on the gate lead exceeds the trigger level.

In many applications, the gate lead of a LASCR will not be used. On some devices, the gate lead doesn't exist at all. If there is a gate lead, it can be used to externally bias the LASCR.

The light intensity required to trigger the LASCR depends on the device's specific design characteristics, and any external biasing (if used).

Chapter 3
Light-Emitting Devices

So far, in Chapters 1 and 2, we have been concentrating on optoelectric devices that detect light energy from some external source, and converts it into some electrical parameter. Such devices are known as *light sensors*. In this chapter we will look at devices that convert electrical energy into light energy. These are light sources, or generators.

INCANDESCENT LIGHTS

Of course, we are all familiar with incandescent lights. An ordinary light bulb is an incandescent device. It converts an electrical voltage into light.

An incandescent light bulb is made from a thin wire in a vacuum. The vacuum is contained within a glass bulb. The wire, which is known as the filament, is usually made of tungsten. When an electrical current is fed through the filament, it becomes very hot and starts to glow. The vacuum and the nature of the tungsten material prevent the filament from burning itself up too rapidly.

The light from an incandescent bulb is usually a mixture of all the colors (white light) unless the bulb, or outside glass envelope, is covered with translucent paint or a gelatin filter of some sort. Light is emitted in all directions. For electronics work, only small incandescent bulbs, like those used in flashlights, are normally used. The schematic symbol for an incandescent lamp is shown in Fig. 3-1.

Fig. 3-1. The simplest light-emitting device is the incandescent lamp.

An incandescent bulb may be driven by either an ac or a dc voltage source. Incandescent lamp bulbs are rated for specific operating voltages. Exceeding this rated voltage will result in premature lamp failure.

Some incandescent bulbs contain a small amount of a special gas (argon, nitrogen, and halogen are typically used) within the glass envelope. The addition of one of these gases tends to extend the filament's operational life and increase brightness. However, the use of one of these gases significantly increases the cost (and sometimes the fragility) of the lamp.

GAS DISCHARGE LAMPS

Another type of lamp bulb also uses an air-tight glass envelope filled with gas. Instead of a filament, the bulb contains a pair of slightly separated electrodes, as illustrated in Fig. 3-2. The most widely used gas for this type of device is neon.

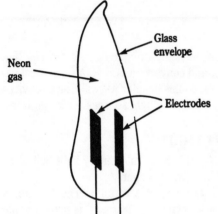

Glass envelope

Neon gas

Electrodes

Fig. 3-2. The neon bulb is an example of a gas discharge lamp.

A voltage is applied across the two electrodes. If this applied voltage is below a specific threshold value, nothing much will happen. When the threshold voltage (generally 60 to 70 volts for most neon lamps), the gas is ionized. This action is also sometimes called *breakdown*. At breakdown, there is an electrical discharge between the two electrodes. The ionized gas glows, so the lamp gives off light. Neon has an orange glow. The schematic symbol for a gas discharge lamp appears in Fig. 3-3.

Other gas discharge lamps use xenon (often used in strobe lights and camera flashes) or mercury vapor.

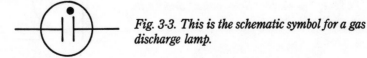

Fig. 3-3. This is the schematic symbol for a gas discharge lamp.

THE LED

For the most part, the optoelectric devices used in this book are semiconductors. A semiconductor light generator is an LED. This is an acronym for light-emitting diode. As the name suggests, an LED is a two-terminal device, made up of a single pn junction. It is used in much the same manner as an ordinary diode.

The two most common schematic symbols for the LED are shown in Fig. 3-4. Which symbol is used is merely a matter of personal preference. The presence or absence of the circle doesn't signify anything in particular. Some technicians simply feel the circle makes the symbol easier to distinguish from the symbol for an ordinary semiconductor diode.

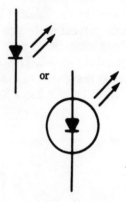

Fig. 3-4. These are the two most commonly used schematic symbols for LEDs.

An LED is a polarized device, since it contains a pn junction. If the LED is reverse-biased, it will act like an ordinary diode, blocking the flow of current. When the LED is forward-biased, it conducts and glows, or emits light.

While some clear (white light) LEDs have been developed, most of these devices emit colored light. Red is by far the most common color for LEDs. Many hobbyists have never even seen an LED that isn't red, but with a little effort you should be able to locate green or yellow LEDs. If you're very lucky, you may even come across an occasional blue LED.

In addition to these visible color types, some LEDs are designed to emit light in the infrared region, which is outside the visible spectrum. Most wireless remote controls use infrared LEDs as their signal source.

LEDs are used primarily as indicator devices. That is, an operator can tell at a glance whether or not a specific voltage is present at a given circuit point by whether or not the LED is lit up.

Within certain limits, the higher the applied voltage, the brighter an LED will glow. Of course, decreasing the applied voltage will cause the LED to glow more dimly. This cannot be used to measure exact values, but it can indicate relative voltage levels.

LEDs are relatively durable devices, but they are intended for use in low-power circuits only. Typically, no more than 3 to 9 volts should be applied to an LED. If the voltage is over 6 volts, an external voltage divider network is highly advisable in most circuits. Excessively high voltages can easily (and quickly) burn out the semiconductor junction, rendering the LED quite useless.

Avoiding excessive voltages is usually easy enough. The circuit designer must always give some consideration to the amount of current flowing through the LED. Excessive current can damage or destroy the component very quickly.

The LED, being a diode, exhibits a very high resistance when it is reverse-biased. According to Ohm's Law (I = E/R), this means that the reverse-biased LED will draw very little current. When forward-biased, on the other hand, the LED's resistance drops to a very low value, allowing the current to climb to a relatively high level. Depending on the rest of the circuitry involved, the LED might well attempt to pass more current than it can safely handle.

The current through an LED must be limited to a safe value. In many circuits, if no external current protection is provided, the LED will draw excessive current and quickly burn itself out. In most LED circuits, a series resistor is used to limit current, as illustrated in Fig. 3-5. The current-limiting resistor will usually have a fairly low value. Generally resistances of about 100 ohms to 1K (1000 ohms) are used for LED current-limiting. Most circuits use a 330-ohm or a 470-ohm resistor for this purpose.

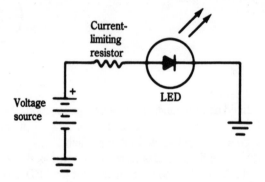

Fig. 3-5. A series resistor is usually required to limit the current flow through an LED.

TRI-COLOR LED

An ordinary LED glows only when it is forward-biased. If it is reverse-biased, it will remain dark. This characteristic allows LEDs to be used to test the polarity of a voltage. Figure 3-6 shows the circuit for an extremely simple polarity checker. The resistor limits the current through both of the LEDs, so only one is needed.

For the most convenient and unambiguous indication of the signal polarity, the two LEDs should be of contrasting colors. For instance, LED1 might be red, and LED2 can be green.

In operation, both the tester and the circuit under test should have a common ground. An alligator clip can be connected to the ground end of the tester. This clip can be fastened easily to any convenient ground point in the circuit under test. A probe is connected to the other end of the tester circuit. This probe is touched to the circuit point you want to test.

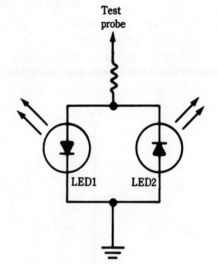

Fig. 3-6. A pair of LEDs can be used to create a simple polarity tester.

If the probe is in contact with an unknown voltage source, you can easily determine the polarity of the unknown signal. If the voltage is positive with respect to ground, LED1 (red) will light up, and LED2 (green) will remain dark. If the signal polarity is reversed, just the opposite will occur; LED2 (green) will glow, and LED1 (red) will be dark. If neither LED lights up, the applied voltage must be zero, or close to it.

What happens if the probe is touched to a circuit point carrying an ac signal? The two LEDs will alternately blink on and off. If the ac frequency is very low, you will be able to see the LEDs blink on and off. At higher frequencies, both LEDs will appear to be continuously lit.

You might be wondering why I am taking the space here to describe this simple circuit. After all, we haven't gotten to the projects yet. This circuit is presented here as an introduction to a specialized variation on the basic LED.

Single-unit dual LEDs are also available. These devices are simply two differently colored LEDs cross-wired and contained within a single package. The LEDs are connected internally as in Fig. 3-6. (Usually the current-limiting resistor is not included. It must be added externally in practical circuits.)

The dual LED will glow one color (usually red) when a dc voltage of one polarity is fed through it. If the polarity is reversed, the dual LED will glow with a second color (usually green). When an ac signal is applied to the device, both internal LEDs glow (actually, they are blinking on and off). Because of the close physical placement and the shared housing, the two colors blend into a single third color (usually yellow). This occurs because the eye actually mixes the two colors to make a third.

As you can see, the dual LED can glow with any of three different colors. For this reason it is known as a *tri-state LED*. They are typically used in more sophisticated indicator applications.

MULTISEGMENT DISPLAYS

LEDs are useful indicators, but they can be even more useful if a number of them are used together to indicate a wider range of circuit conditions.

Figure 3-7 shows a simple demonstration circuit in which three LEDs indicate the position of a rotary switch. For convenience, the current-dropping resistors are not shown here, though they should be included in a practical circuit.

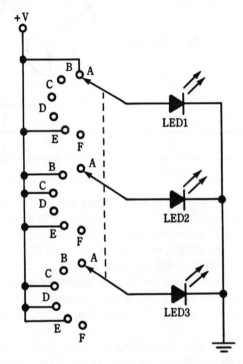

Fig. 3-7. This simple demonstration circuit shows how multiplexed LEDs can indicate the position of a rotary switch.

In a practical circuit, this switch would probably have additional poles that simultaneously perform other functions. For our purposes here, we are only concerned with the LED indicators.

If the switch is in position A, only LED1 will be lit. LED2 and LED3 will remain dark. Advancing the switch to position B will extinguish LED1 and LED2 will light up. LED3 will still stay dark. If the switch is moved to position C, both LED2 and LED3 will be lit, while LED1 will be dark. If only LED3 is lit and the other two LEDs are dark, we know the switch must be in position D. Position E illuminates all three LEDs, while in position F, all three LEDs are turned off.

The operation of this circuit can be summarized as follows:

Switch	LED1	LED2	LED3
A	O	X	X
B	X	O	X
C	X	O	O
D	X	X	O
E	O	O	O
F	X	X	X

In this summary chart, an O means that the indicated LED is on, or lit. An X indicates that the LED in that column is off, or dark.

Study this circuit diagram carefully. Make sure that you thoroughly understand how the three LEDs are being controlled in each of the six switch positions. As an extra exercise, redesign the circuit to add a seventh switch position (G). In this position, LED1 and LED3 should be lit, while LED2 remains dark.

Note that the cathodes of all three LEDs in Fig. 3-7 are electrically tied together. In effect, all three LEDs share a single cathode. Such a system is called a *common-cathode display*. Alternatively, the LEDs could be connected as a common-anode display. This is illustrated in Fig. 3-8. Note that all we've done here is reverse the polarities of the LEDs and the power supply.

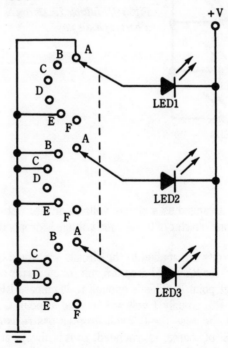

Fig. 3-8. The LEDs could be connected in a common-anode arrangement, rather than the common-cathode system shown in Fig. 3-7.

Note that in these cases the word "common" does not necessarily refer to the circuit's common ground point. It simply refers to a shared element that is common to each of the component LEDs.

Bargraphs

Figure 3-9 shows another simple, multiple-LED display format. In this circuit, the higher the applied voltage is, the more LEDs that will light up. For example, let's assume that each LED requires at least 1.5 volts to light, and that each resistor has a value of 1000 ohms (1K). For simplicity we will ignore the internal resistances of the LEDs themselves.

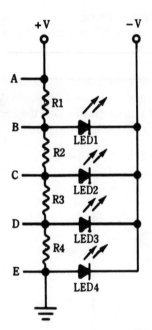

Fig. 3-9. Multiple LEDs can be arranged to create a bargraph display.

The resistors are arranged as a simple voltage divider network. Because all four resistances are equal, one quarter of the applied voltage is dropped across each individual resistor.

For instance, if 2 volts are applied to the circuit, we would be able to read the full 2 volts at point A. Resistor R1 would drop 0.5 volt (one quarter of the source voltage— ¼), leaving 1.5 volts at point B. This is enough to illuminate LED1. Another 0.5 volt is dropped by resistor R2, so only 1 volt will be read at point C. This means LED2 is off. LED3 is also dark, because the 0.5-volt drop across resistor R3 leaves only 0.5 volt at point D. Point E, of course, is grounded, so it is always at zero potential. LED4 will never light up. It is included here for illustrative purposes. The fourth LED would be eliminated in a practical circuit, since it doesn't serve any real purpose.

Now, what happens if we increase the source voltage to 4 volts? In this case, 1 volt will be dropped across each individual resistor (⁴⁄₄ = 1). Now we will have the following voltages at the test points;

A	4 volts
B	3 volts
C	2 volts
D	1 volt
E	0 volts

Because points B and C are greater than the minimum LED turn-on voltage (1.5 volts), LED1 and LED2 will light up, while LED3 (and LED4, of course) will remain dark. Raising the input voltage to 6 volts will result in a 1.5-volt drop across each resistor;

A	6 volts
B	4.5 volts
C	3 volts
D	1.5 volts
E	0 volts

In this case LED1, LED2, and LED3 will be lit.

Naturally, this type of circuitry can be expanded readily to include more than just three active LEDs. However, recall that there is an inherent limitation to how much voltage can be applied safely to an LED. To measure higher level signals, some sort of attenuation stage is necessary.

Most practical circuits of this type use comparators to control the LEDs. For some practical projects of this type, see Chapter 9.

This kind of display is often called a bargraph, because the LEDs are normally arranged as a line, or bar, as illustrated in Fig. 3-10. The longer the lighted portion of the bar, the greater the level of the input signal.

Fig. 3-10. The longer the illuminated "bar," the higher the input signal is.

Bargraphs can be used to indicate voltage, current, or any other electrical parameter. In many modern cassette recorder decks, LED bargraphs are used for the VU meters.

A variation on the bargraph is the dot-graph. In this case, only the highest appropriate LED is lit. All lower (and higher) LEDs stay dark. A typical dot graph display is illustrated in Fig. 3-11.

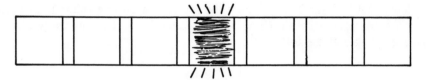

Fig. 3-11. A variation on the bargraph display is the dot-graph display.

Bargraph and dot-graph displays are very easy to read, even at a distance or at an angle. They are not as precise as a panel meter or a digital read-out, but in many applications a bargraph or dot-graph display is sufficiently accurate.

Some manufactures sell strips of LEDs in bargraph form. This is just a number of LEDs physically joined together, usually with squared off lenses to increase the effect of a continuous line.

Seven-Segment Displays

Perhaps the most widely used multi-LED display arrangement is the seven-segment display. This consists of seven LEDs in a single housing. The LEDs are shaped like narrow rectangles and arranged in a figure-eight pattern, as shown in Fig. 3-12. Most seven-segment displays actually include an eighth LED that is used as a decimal point. We can ignore this extra LED here.

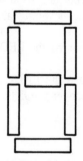

Fig. 3-12. A seven-segment display contains seven rectangular LEDs arranged in a figure-eight pattern.

A seven-segment display may be of either the common-cathode or the common-anode type. There is only a single lead for the common element. There other (active) leads for the individual LED segments are brought out individually.

If all seven LED segments are lit, the display will look like the digit 8, of course. But by lighting only selected segments, any single digit from 0 to 9 can be formed. For example, if segments A, B, D, E, and G are lit, but segments C and F are dark, the numeral 2 will be formed.

Each of the ten digits in the decimal system can be formed by choosing the appropriate segments. How each digit is formed is illustrated in Fig. 3-13 through 3-22. Of course, if a number higher than 9 must be displayed, more than one seven-segment display must be used.

Fig. 3-13. The numeral "0" as displayed on a
seven-segment display unit.

Fig. 3-14. The numeral "1" as displayed on a
seven-segment display unit.

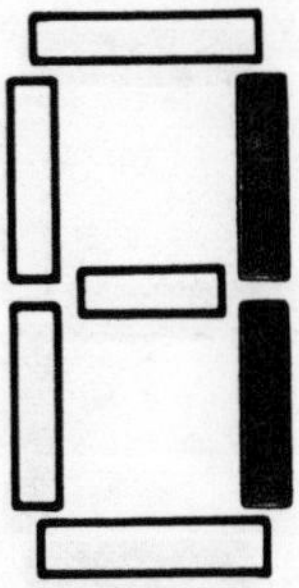

Fig. 3-15. The numeral "2" as displayed on a
seven-segment display unit.

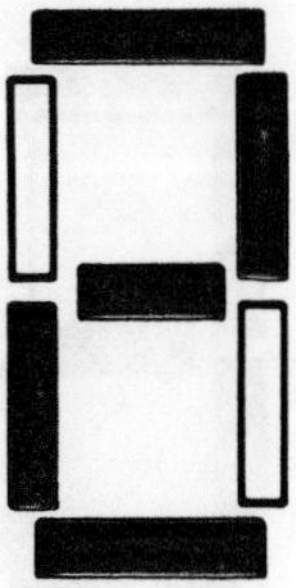

Fig. 3-16. The numeral "3" as displayed on a
seven-segment display unit.

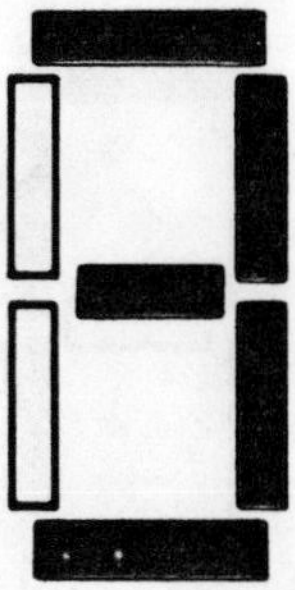

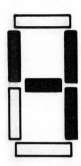

Fig. 3-17. The numeral "4" as displayed on a seven-segment display unit.

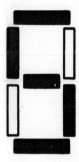

Fig. 3-18. The numeral "5" as displayed on a seven-segment display unit.

Fig. 3-19. The numeral "6" as displayed on a seven-segment display unit.

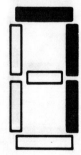

Fig. 3-20. The numeral "7" as displayed on a seven-segment display unit.

Fig. 3-21. The numeral "8" as displayed on a
seven-segment display unit.

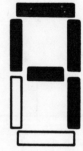

Fig. 3-22. The numeral "9" as displayed on a
seven-segment display unit.

Certain letters of the alphabet can also be displayed on a seven-segment display, although some occasionally require a little imagination to read. Can you determine which segments must be lit to display each of the following letters?;

A, B, C, E, F, G, H,
I, L, O, P, S, U

Can you find a way to display any other letters or symbols with a seven-segment display?

For most seven-segment displays, an external current-limiting resistor must be used for each individual segment. Some display-driver ICs feature built-in current-limiting, so external resistors are not required.

Figure 3-23 shows how a common-cathode seven-segment display usually appears in schematic diagrams. The symbol for a common anode seven-segment display is very similar, as shown in Fig. 3-24. If there is no indication of which type of display is being used, just look to see where the common element is connected. If it is connected to ground (or a negative voltage, or a very low, positive voltage), it is a common-cathode device. If the common element is connected to the V+ line, or some other relatively high positive voltage, the display unit is a common-anode type.

More complex multisegment displays are available. Some can display any letter of the alphabet and many common symbols. These devices are often awkward to use. They are also very difficult to locate (especially on the hobbyist level), and quite expensive if you do find them. For these reasons they are not generally used in hobbyist projects.

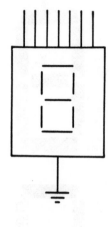

Fig. 3-23. This is how a common-cathode seven-segment LED display is usually shown in schematic diagrams.

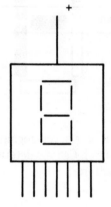

Fig. 3-24. This is how a common-anode seven-segment LED display is usually shown in schematic diagrams.

INFRARED DEVICES

We ordinarily think of light as being visible. In a way, "invisible light" sounds like it might be an oxymoron, but in fact, visible light is only a small part of a large electromagnetic spectrum. We can only see a small band of frequencies. The specific frequency determines the color.

Light actually extends far beyond either end of the visible spectrum. Light with a frequency in the region just above the visible spectrum is called *ultraviolet*. Light with a frequency in the region just below the visible spectrum is called *infrared*.

Infrared light is often used in optoelectric devices, especially in applications where a visible beam of light would be a disadvantage. Infrared applications include wireless communication, security systems, wireless remote control, and related functions.

Infrared LEDs are widely available. They are quite similar to ordinary LEDs, both in their construction and their use. The only real difference is the frequency of the light energy emitted by the device. The same sort of circuitry can be used for both standard LEDs and infrared LEDs, though, of course, the actual applications are likely to be different. Because infrared light is invisible to the human eye, an infrared LED wouldn't be of much use in the indicator applications that are most common for ordinary visible light LEDs.

Special sensors must be used to detect the invisible light generated by an infrared LED. You can't tell if the infrared LED is working or not, just by looking at it. A number of infrared sensor devices are available. They are basically similar to the light sensor devices described in Chapters 1 and 2, except they are designed to be sensitive to light energy with a frequency in the infrared region.

FLASHER LED

An interesting variation on the basic LED is the flasher LED. It might be a little difficult to tell the difference at first glance. A flasher LED is generally a little larger than most standard LEDs. However, standard LEDs do come in many sizes, and some are as large as flasher LEDs.

If you look very closely at a flasher LED, you will see a small black speck within the clear epoxy housing. This speck is actually a tiny oscillator IC (integrated circuit). When a voltage is applied to the two leads of this device, the LED will blink on and off at a rate of approximately 3 Hz. (The exact flash rate is determined by the IC's internal design.)

The schematic symbol for a flasher LED is shown in Fig. 3-25. This is close to the symbol for a standard LED, so always be careful. The flasher function is indicated by the two-tier cathode line.

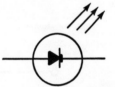

Fig. 3-25. A flasher LED has a tiny built-in oscillator IC.

The flasher IC is very easy to use, because virtually everything except the power supply is built into the LED housing itself.

The internal oscillator IC even includes its own internal current-limiting resistor, so even this component doesn't have to be added externally. The IC is designed for use with a +5-volt dc power supply and draws about 200 mA. Higher voltages may be applied to the flasher LED if an external series-voltage-dropping resistor is used, as illustrated in Fig. 3-26. For a +9-volt power supply, the dropping resistor should have a value of 1000 ohms (1K) at ½ watt.

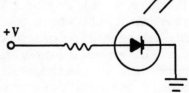

Fig. 3-26. Higher voltages can be applied to a +V *flasher LED if an external series voltage dropping resistor is used.*

The flasher LED can even be run off of an ac power source by placing a diode in parallel across the LED, as shown in Fig. 3-27.

The flash rate can be speeded up easily just by adding a capacitor in parallel across the dropping resistor, as shown in Fig. 3-28. You will probably have to experiment with various component values to obtain the desired flash rate. Generally, capacitance values should be kept in the 500 μF to 3000 μF range for most practical applications. Flash rates above about 10 to 12 Hz tend to blend together as seen by the human eye. If the flash rate is higher than this, the LED will simply appear to be continuously lit.

More sophisticated flasher circuits (using external circuitry with standard LEDs) are featured in Chapter 8.

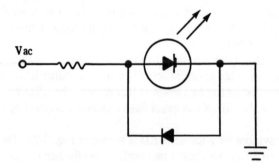

Fig. 3-27. A flasher LED can be powered by an ac voltage source by using a diode in parallel.

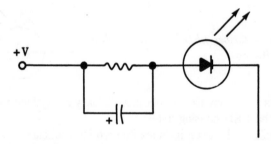

Fig. 3-28. An external capacitor can be used to increase the flash rate of a flasher LED.

USING LEDs AS LIGHT SENSORS

LEDs ordinarily are used to generate light, but they can also be used to detect the presence and approximate level of an external light source. Remember, all semiconductors are photosensitive, at least to some degree. An LED must be enclosed in a transparent case to let its generated light get out. The same clear housing also lets external light get in and strike the LED's pn junction. As a result, the LED can be used as a "quick and dirty" light sensor if a dedicated device doesn't happen to be available at the moment.

If an LED is exposed to a strong external light source, a small voltage is generated between its leads. The exact magnitude of this voltage is determined by the intensity (brightness) of the external light source and the actual physical structure of the LED itself.

LED light detectors are most sensitive to the type and frequency of light they are designed to emit themselves. For instance, a red LED will respond best to red light, a green LED will be more sensitive to green light, and so forth.

LASER DIODES

By modifying the basic design of an LED, a very powerful device can be created. This modified LED-type device can generate a very coherent, monochromatic beam. "Coherent" in this sense means that all of the light energy is in phase and focused in a narrow, very intense beam. *Monochromatic* means that the light produced is composed of a very narrow range of frequencies, or in other words, is a pure, single color.

Some readers might recognize this description. Such a coherent, monochromatic beam of light is known as a laser beam. The modified LED device is a *laser diode*. More precisely, this type of device is known as an *injection diode*.

Laser diodes will not be used in any of the projects in this book, even though they certainly qualify as optoelectric devices. Even though we won't be using laser diodes here, it may be useful to have a rough understanding of the general theory behind them. The explanation here will be very simple. If you are particularly interested in this area, you should read a book specifically on lasers, such as *Lasers: The Light Fantastic* (TAB book #2905) by Clayton L. Hallmark and Delton T. Horn.

The injection laser is very similar to the basic LED. When the applied voltage across an LED's leads exceeds a certain threshold value, the LED starts to glow. If there is no applied voltage, or if the applied voltage is below this threshold value, the LED doesn't do much of anything. It remains dark. Figure 3-29 illustrates the internal physical structure of an ordinary LED.

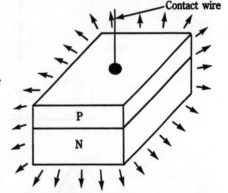

Fig. 3-29. This is the internal physical structure of an ordinary LED.

In a laser diode, there is a second critical threshold point that is significantly higher than the LED's turn-on voltage. This second threshold point is generally called *Jth* or *Ith* in the technical literature.

If the applied current is below the Jth threshold, the laser diode functions exactly like an ordinary LED. It glows with a relatively broad spectrum of wavelengths. The light is emitted in a wide pattern of radiation. No lasing action takes place.

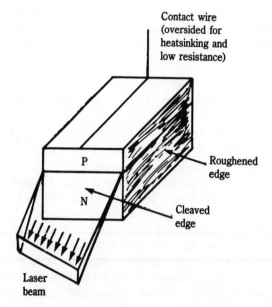

Fig. 3-30. This is the internal physical structure of a laser diode.

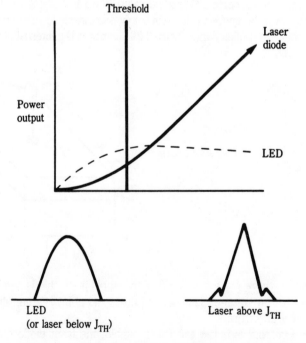

Fig. 3-31. The operation of an LED is compared with that of a laser diode.

When the current applied to an injection laser exceeds the Jth threshold level, the emitted light narrows down to a very thin beam, and lasing occurs within the device. The emitted light becomes highly coherent and monochromatic, and escapes from both of the diode's end facets unless one is coated with a reflective (usually gold) film.

The internal physical structure of an injection laser diode is illustrated in Fig. 3-30. Compare this diagram to the drawing of the basic LED in Fig. 3-29. The functioning of an injection laser diode and a standard LED are compared in Fig. 3-31.

LCDs

LEDs are handy for many indicator and display applications. They are small, inexpensive, sturdy, and quite easy to use.

However, there are disadvantages to using LEDs. They are often difficult to see when the ambient lighting is bright. (Sunlight particularly tends to wash out the glow of LEDs.) In addition, LEDs tend to eat up a considerable amount of current, especially in circuits using multiple LEDs. This is often a problem in battery-operated circuits where multiple-digit seven-segment displays are utilized. An increasingly available alternative is to use an LCD (Liquid Crystal Display) in place of the LED(s).

Strictly speaking, LCDs are not really optoelectric devices, at least in the sense the term is being used in this book. We will not be using LCDs in any of the projects. Still, it is worthwhile to know something about these increasingly important devices. They are certainly close related to the topic at hand.

An LCD display panel is an optically transparent "sandwich," usually with an opaque backing of some kind. The inner faces of the two panels that make up the sandwich are covered with a thin metallic film. On one of the panels, this film coating has been deposited in the form of the desired display. A common format is similar to the seven-segment display discussed earlier. Letters and symbols can also be used.

Between the two panels is a special fluid called a *nematic liquid*. Under ordinary circumstances, this fluid is transparent. However, when an electrical field is passed between the back metalized panel and one of the metalized segments on the front panel, the liquid between these sections will darken and become opaque. The segment will appear as a black mark. When the electrical field is removed, the nematic liquid becomes transparent again.

A very small amount of current is required to darken an LCD segment. The LCD does not generate light. Some external light source must illuminate the display in order for anyone to read it. The segments are easily visible under most ambient lighting conditions, except when the lighting is very dim. Many LCD units feature a built-in backlight of some kind to overcome any problems in dimly lit environments.

LCDs can be designed to display almost anything, including numerals, letters, symbols, and even simple drawings.

Chapter 4

Optoisolators
and
Fiberoptics

This chapter covers optoelectric devices that are not exactly either sensors or generators. Well, what are they then? Just because something is not easily classifiable, that certainly doesn't mean that it isn't a very useful and versatile device.

The first part of this chapter covers optoisolators. This type of component contains both a light sensor and a light generator in a single package. The idea might sound rather pointless at first, but optoisolators have countless practical applications.

The second part of this chapter discusses fiberoptic cables, which are a transmission medium for light-wave signals.

OPTOISOLATORS

If we connect an LED (light generator) and a photoresistor (light sensor) as shown in Fig. 4-1, the signal applied to circuit A will be transferred to circuit B without a direct electrical connection between the two circuits.

The signal applied across the LED controls its brightness from instant to instant. The photoresistor is positioned to detect the light from the LED. The photoresistor's resistance at any given instant is determined by the current brightness of the LED. The original input signal from circuit A is converted to a varying resistance in circuit B.

The only connection between circuits A and B is through the light passing from the LED to the photoresistor. The two circuits are electrically isolated. This type of system is called an *optoisolator*. Sometimes the name *optocoupler* is used. For all intents and purposes, the two terms are interchangeable.

Figure 4-1 shows a very rough optoisolator system. One problem with the system as shown here is that there can be considerable interference from external (uncontrolled) light sources. To prevent such interference, the LED and the photoresistor should be enclosed in a common housing that is as light-tight as possible. The interior should be

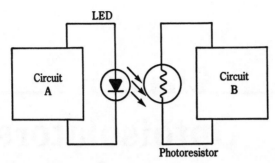

Fig. 4-1. A light generator can transfer signals to a light sensor without an electrical connection between the two circuits.

very nonreflective. In a home-made optoisolator, the interior of the case should be painted flat black to eliminate reflections.

Relative placement of the LED and the photoresistor is fairly critical. Light intensity drops off sharply with increased distance, so even changing the relative placement of the components by a fraction of an inch could totally change the response of the optically coupled circuit.

For maximum efficiency in the signal transfer, the LED should be precisely positioned so it sheds as much light as possible onto the sensor.

Fortunately, commercial optoisolators are widely available, so you don't have to worry about making your own. The only real reason to build a home-brew optoisolator is for experimenting and to learn how such a system works.

A commercial optoisolator looks much like an ordinary IC, but instead of an integrated circuit, the DIP housing contains an LED and a light sensor. The epoxy case is completely light-tight, eliminating external interference problems and maximizing coupling efficiency. There is no place for the light energy produced by the internal LED to go except to the sensor.

The sensor does not necessarily have to be a photoresistor, as in our example. Optoisolators with any of the light sensors discussed in Chapter 2 are available, including:

- photodiodes
- photodiacs (bidirectional diodes)
- phototransistors
- photoFETs
- LASCRs (light-activated SCRs)
- photoDarlingtons
- high-voltage phototransistors
- photothyristors

and other active light-sensing devices.

Active sensor devices, such as phototransistors, respond much more quickly than simple photoresistor-type sensors. Phototransistors seem to be the most commonly used sensors for commercial optoisolators today.

Some optoisolators use visible light, while others use infrared light. Because the light is entirely contained within the device's housing, the difference generally isn't of much importance to the hobbyist or experimenter.

Within the DIP package, a thin sheet of optical glass is placed between the LED and the photosensor. This glass barrier is designed to maximize the optical transfer between the two internal photocomponents. In addition, this separating sheet of glass improves the electrical isolation between the optoisolator's input and output.

Commercial optoisolators offer extremely high isolation between the input (LED circuit) and the output (sensor circuit). Isolation ratings of up to 2500 volts are not uncommon.

Optoisolators are often used for safety reasons—both for the protection of human operators and for the protection of sensitive circuit components. A high-voltage ac circuit can be safely coupled to a low-voltage dc circuit, for example.

Optocoupling is less prone to interference pickup than most other coupling methods. A good optoisolator is considerably less bulky (and exhibits less signal loss) than an isolating (or coupling) transformer, which was once the most common approach to coupling signals from one circuit to another.

Optocoupling can also result in fewer loading problems on the source circuit.

In schematic diagrams, optoisolators are sometimes drawn as "black box" ICs. That is, the component is represented as a simple box, with numbered lines coming off to represent the individual leads.

More commonly, the internal LED and photosensor are drawn in the schematic, as shown in Fig. 4-2. The dotted box around these two items indicates that they are contained within a single housing. That is, an optoisolator is being shown, not two discrete devices. Pin numbers for the optoisolator leads are usually given.

Fig. 4-2. This is the schematic symbol for a typical optoisolator.

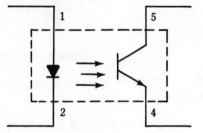

FIBEROPTICS

In one way or another, most optoelectric applications are a form of communication. For example, an optoisolator permits one-way communication from one circuit to another.

Whenever optical communications are used, there are three inherent problems that must be dealt with. These problems can be defined simply as interference, dispersion, and line-of-sight limitations.

Interference is caused by any (uncontrolled) external light source. To the sensor, light is light. It doesn't care where the light comes from. Interference can be reduced in some applications by using a sensor that is sensitive only to a limited portion of the light spectrum. (A filter can be placed over the sensor to achieve the same effect.) Unfortunately for our purposes here, white light contains all light frequencies. Therefore, there will still be some interference even if a frequency-limited sensor is used.

In an optoisolator, the interference problem is side-stepped by enclosing the desired (controlled) light source and the sensor in a single light-tight enclosure. No external light can get in to interfere with the desired signal. Obviously, this solution is not applicable to all optoelectric applications. Usually the desired light source and the sensor are too far apart to share a common housing. Besides, in many applications it is necessary for something to pass between the light source and the sensor. Quite plainly, a light-tight housing is out of the question in such applications.

Optoelectrical systems also have an inherent problem with dispersion. Light from most sources (lasers are an obvious exception) disperses as it moves out from its source. Even if the light starts out as a directional beam, the beam usually widens dramatically as it moves further from the source. This is illustrated in Fig. 4-3.

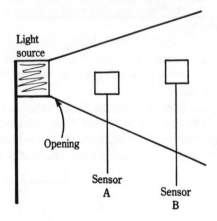

Sensor B receives much less energy than Sensor A

Fig. 4-3. Ordinary light sources suffer from dispersion that reduces the energy as the distance increases.

For a sensor of a given size, placing the sensor further away from the source reduces the amount of light energy striking the sensor. At some point, the light energy becomes too weak to be detected reliably by the sensor.

In an optoisolator, this is not a problem. The light source and the sensor are physically very close to one another. In addition, the light-tight housing that keeps interfering external light out also keeps the light from the source in. There isn't anyplace for the light energy to go except to the sensor.

Once again, we can't use the optoisolator solution for many optoelectrical applications. The more focused the source light beam is, the less energy that will be wasted.

Of course, light-beam communications are always limited to a line-of-sight path. Any opaque object in the path between the light source and the sensor will block off the communicating light beam. In some applications, such as object counters and intrusion detectors, this blocking of the light beam is the whole point of the system. In other applications, the line-of-sight limitation is a major problem. Of course, there is no line-of-sight problem for optoisolators. The light source and the sensor share an enclosed housing, so nothing can ever get between them and block the flow of light from the source to the sensor.

In some applications, a great way to limit interference, dispersion, and line-of-sight problems is to use a fiberoptic cable between the source and the sensor. The fiberoptic cable is made up of a number of thin transparent glass or plastic fibers. Some of these fibers are no thicker than a single human hair. A fiberoptic cable functions rather like an ordinary copper cable (or wire), except light waves are being conducted rather than electrons.

Ordinarily light waves travel only in straight-line paths. A reflective object, such as a mirror, will cause the light beam to change direction and adopt a new straight-line path. Most objects, however, will block any further travel of the light beam. Light waves cannot normally go around corners or follow a curved path.

In an optical fiber cable, the light waves can be forced to follow any curve or any sharp angle or bend in the cable. This effect is due to refraction. The refractive index if the fiber's core is slightly higher than that of its cladding. If you are not familiar with the physics of light, the last sentence might not mean very much to you. That's okay; we don't need to go into refraction theory here except to say that differing refractive indexes cause light beams to be bent. In a fiberoptic cable, the light is effectively confined to the interior of the fiber, regardless of any bends or curves in the cable. This effect is illustrated in Fig. 4-4.

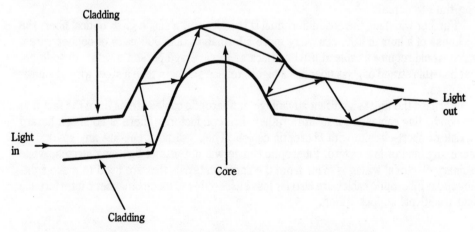

Fig. 4-4. Refraction keeps almost all of the light energy contained within an optical fiber.

A transmitter or light source (usually modulated in some way) is placed at one end of the fiberoptic cable. At the other end we have a receiver, or photosensitive sensor. Special linkage devices are available to ensure the best and most secure mechanical and optical connection possible. In operation, the effect is similar to placing the source and the sensor side-by-side. There is a limit to the length of the fiberoptic cable, of course. Over long runs, the signal reaching the sensor may be significantly reduced. Such losses generally won't be noticeable in fiberoptic cables only a few yards or so long. For really long connections (such as in telephone lines), intermediate amplifier stages are placed at strategic points along the length of the fiberoptic cable. Similar amplification stages are needed in long lengths of ordinary electrical cables, too. If a laser light beam is used as the signal source, a typical fiberoptic communications system can use connecting lines as long as ten kilometers without any intermediate amplification stages.

Incidentally, for long-line communications systems, another important consideration is that a fiberoptic system is inherently difficult to jam or intercept without authorization.

An ordinary light source may be used as the transmitter for a fiberoptic communications system, but laser light sources are being increasingly employed in such systems. The intense power and coherence of laser light gives increased range, and more data can be transmitted simultaneously on a single fiberoptic cable.

Optical fiber cables can be used in almost any application where an electrical conductor cable would normally be used. In a sense, a fiberoptic system could be considered a long-distance optoisolator. Fiberoptic cables offer several advantages over standard electrical conductors in many applications.

For one thing, fiberoptic cables tend to be cheaper than traditional coaxial cables, and there is no reason to assume that this price difference won't be even greater in the future. Coaxial cables are normally made of copper wires and the dielectric is made from petroleum-derived plastics. The raw materials for coaxial cables are inherently expensive. The basic raw material for glass fiberoptic cables, on the other hand, is simple, ordinary sand. That is an ingredient that is likely to remain plentiful and cheap for a long, long time to come.

Further widening the cost differential is the fact that a single glass optical fiber, the thickness of a human hair, can carry more information than 900 pairs of copper wires, which would require a cable at least as thick as the average person's wrist. In addition, that hair-thin strand of glass fiber has a higher tensile strength than a steel wire of similar thickness.

Perhaps the most significant advantage of fiberoptic cables stems from the fact that the conducting cables are insulators rather than conductors. There is no shock hazard or risk of short-circuits with fiberoptic cables. They will not generate any sparks, or cause any kind of fire hazard. Fiberoptic cables will not attract lightning strikes, while ordinary electrical wiring is made from the same materials that are used to make lightning rods. Fiberoptic cables are also far less susceptible to electromagnetic interference than traditional copper wiring.

Part II
PROJECTS

Chapter 5

Power Supply Projects

The projects in this chapter are all pretty straightforward. Because they are all related to power supplies, these projects are intended to be used in conjunction with other projects. They aren't particularly exciting in and of themselves, but they can be a useful part of many electronics systems.

PROJECT 1: SOLAR-BATTERY PACK

Battery power permits almost any electronic circuit to become portable. To operate the circuit, you don't have to be anywhere near an ac outlet. In many cases, even when you are operating the equipment indoors, finding a free electrical socket can be a nuisance.

Batteries are not ideal, however. (What is?) They add to the bulk and weight of the project. Individual batteries (or cells) aren't terribly expensive. (At least, not usually. Some uncommon battery types carry pretty hefty price tags.) In the long run, however, battery power proves to be quite expensive in almost all cases, unless the current drawn by the circuit is extremely low. Batteries wear out and have to be replaced. Heavy current loads can require very frequent replacements. Batteries often tend to go dead at the most critical and inopportune times. (This is in accordance with Murphy's Law, of course. If anything can go wrong, it will.)

In many applications, rechargeable batteries may be a good choice. In some applications however, this solution doesn't offer much of an improvement. When the charge wears off, it often happens very quickly, with very little warning. It is a time-consuming nuisance to recharge the batteries. (Typically, recharging takes about 8 hours, or so.) Rechargeable batteries tend to be rather expensive. Many types of rechargeable batteries, such as Ni-Cads (nickel-cadmium cells) put out a lower than normal voltage (1.25 volts instead of 1.5 volts per cell), which can be a problem with certain devices. Rechargeable batteries can die completely from time to time. This usually happens very unexpectedly.

Solar power is almost like getting a power source for nothing. The photovoltaic cells will cost more than batteries (even rechargeable batteries), but it is a one-time expense. As long as the photocells are not damaged (they are delicate), they won't need replacement.

A photovoltaic cell converts light energy into electrical energy that can be used to power an electronic circuit. While this is often called "solar power," it does not require sunlight. Any type of light source will do fine. Figure 5-1 shows a general-purpose solar-battery pack. A parts list is given in Table 5-1.

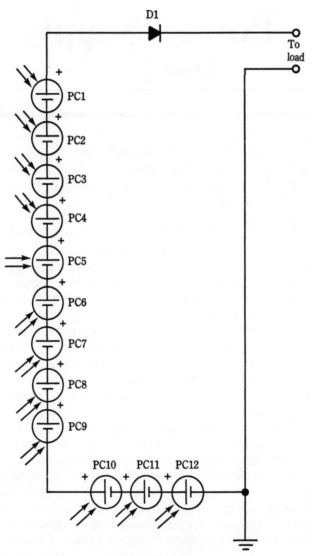

Fig. 5-1. Project 1: general-purpose solar-battery pack.

Table 5-1. Parts List for the
General-Purpose Solar-Battery-Pack Project of Fig. 5-1.

Component	Description
PC1–PC12	photovoltaic cell—silicon (Radio Shack #276–124, or similar)
D1	1N914 diode, or similar

This circuit is quite simple. Twelve photovoltaic cells are shown here. You can vary this number to get the desired supply voltage. There is an upper limit on how much voltage can be supplied in this manner. Each cell produces approximately 0.5 volt. (The photovoltaic cells specified in the parts list are rated for 0.42 volt per cell.) Extra surface space is required for each additional cell. Moreover, the cells must be mounted so that they all are more or less evenly illuminated.

The parts list calls for the 276-124 cell, which is available at Radio Shack stores nationwide. You can use other cells from other sources, if you prefer. Surplus houses often offer great bargains.

The Radio Shack cells measure 2.5 cm by 5 cm. If you arrange the twelve cells as shown in Fig. 5-2, the entire array will measure 10 cm by 15 cm. You can see how the array size can rapidly grow very unwieldy as the output voltage is increased.

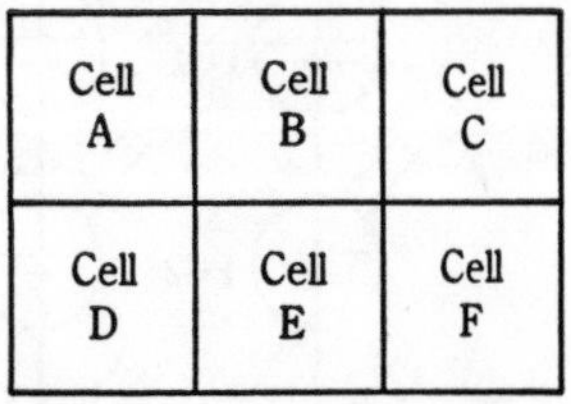

Fig. 5-2. By arranging the cells as shown here, the solar-battery pack can be as small as 10 cm by 15 cm.

Cell A	Cell B	Cell C
Cell D	Cell E	Cell F

Remember that photovoltaic cells are physically fragile. The array should be securely mounted on a firm, sturdy base to minimize the chances of accidental damage. Be careful never to drop your solar-battery pack, especially onto a hard surface. You might be lucky. Then again, you might not.

The battery array should be mounted to receive the maximum possible light energy. All ratings for photovoltaic cells are maximums. If the cells are not adequately lit, they will not put out the rated power levels.

The current available to the circuit is limited by the specific photovoltaic cells used. It will never be very high. A solar-battery pack like this is only suitable for powering very low current loads. The cells mentioned in the parts list are rated for 0.2 amp.

All of the cells in the battery pack must be identical. Do not intermingle cells with different ratings. The lower current cells can soon be destroyed. The total current output of the battery pack as a whole is equal to the current rating of any individual cell in the array.

The diode is somewhat optional. Its purpose is to prevent any back-voltage from feeding back into the photovoltaic cells. A voltage applied across a photovoltaic cell (especially with reverse polarity) can easily damage the delicate semiconductor crystal. The diode permits current to flow from the photovoltaic cells into the load (circuit), but not in the opposite direction. In some applications, this diode will be unnecessary. Some circuits do not present any risk of feeding back any power. However, I believe in "better safe than sorry," and a diode is pretty cheap insurance, all things considered.

PROJECT 2:   HIGH-CURRENT SOLAR-BATTERY PACK

If you want to use solar power with a project that requires more current than your photovoltaic cells can provide, this high-current solar-battery pack project might be just what you need. The circuit is shown in Fig. 5-3, with a parts list appearing in Table 5-2. You can substitute different photovoltaic cells, if you prefer. Be sure to check the manufacturer's current and power ratings for the device you use. All of the cells in your battery pack should be electrically equivalent.

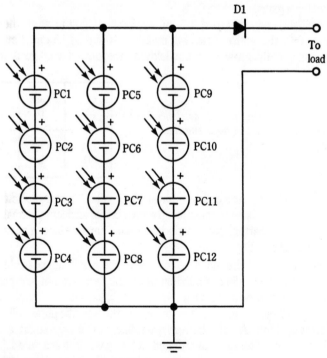

Fig. 5-3. Project 2: high-current solar-battery pack.

**Table 5-2. Parts List for the
High-Current Solar-Battery-Pack Project of Fig. 5-3.**

Component	Description
PC1–PC12	photovoltaic cell—silicon (Radio Shack #276–124, or similar)
D1	1N914 diode, or similar

The available output current is increased by connecting multiple solar batteries in parallel. For the arrangement shown here, each sub-battery contains four cells. Using the photovoltaic cells specified in the parts list, each cell puts out 0.42 volt at 0.2 amp. Each sub-battery has a total voltage of

$$4 \times 0.42 = 1.68 \text{ volts}$$

or approximately 1.5 volts. The current rating for each sub-battery is 0.2 amp.

Because the three sub-batteries are in parallel, the output voltage for the total combination remains at about 1.5 volts (1.68 volts). However, the current capabilities of each sub-battery add for a higher total effective rating. The parallel combination can handle currents up to

$$0.2 + 0.2 + 0.2 = 0.6 \text{ amps}$$

As with all solar-battery arrays, there is a practical limit to how much voltage and current can be supplied. The surface area of each cell must be mounted for exposure to light. All of the cells in the array must receive more or less equal illumination. It is very easy to reach a point where the physical size of the array will simply be too unwieldy to be of any practical use.

Remember that photovoltaic cells are physically fragile. The array should be securely mounted on a firm, sturdy base to minimize the chances of accidental damage. Always be careful never to drop your solar-battery pack.

As in Project 1, the diode is somewhat optional. Its purpose is to prevent any back-voltage from feeding back into the photovoltaic cells. A voltage applied across a photovoltaic cell (especially with reverse polarity) can easily damage the delicate semiconductor crystal. The protective diode might not be necessary in all applications, but it is cheap insurance. Why not include the diode, just in case?

PROJECT 3: SOLAR-BATTERY RECHARGER

If rechargeable batteries are more suitable for your project than either of the solar-battery packs in projects 1 and 2, then you can use photovoltaic cells to build a solar-powered

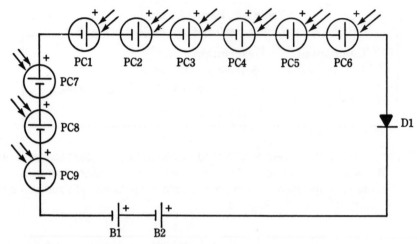

Fig. 5-4. Project 3: solar-battery recharger.

Table 5-3. Parts List for the
Solar-Battery Recharger Project of Fig. 5-4.

Component	Description
PC1–PC9	photovoltaic cell—silicon (Radio Shack #276–124, or similar)
D1	1N914 diode, or similar **DO NOT OMIT!**
B1, B2	nickel-cadmium cells to be recharged

battery recharger. This circuit is intended for use with trickle-charge (slow charge) Ni-Cad (nickel-cadmium) batteries, and is shown in Fig. 5-4. Table 5-3 give a typical parts list for this project.

Basically, there isn't much difference between this circuit and a solar-battery pack (Project 1). Nine photovoltaic cells generate 3.78 volts, which is sufficient to trickle charge two Ni-Cad cells (rated for 1.25 volts each). Note that the solar-battery voltage is slightly higher than the rated Ni-Cad voltage (2.50 volts). This helps force the charge into the Ni-Cad cells. For best results, the Ni-Cad cells should be fully discharged before you attempt recharging.

In use, the current supplied by the photovoltaic cells must not be permitted to exceed the maximum safe charging current for the Ni-Cad cells being recharged. If too much charging current is used, the Ni-Cad cells may be irreparably damaged.

As in the preceding projects, the diode (D1) prevents a back voltage from feeding back into the solar cells. In this project, this diode is absolutely essential, so do not omit this component. If the diode is not present in the circuit, the Ni-Cads will attempt to discharge through the photovoltaic cells, especially during periods of darkness or less than full lighting. Even a passing shadow could cause the Ni-Cads to start discharging. Besides defeating the purpose of the battery recharger, this could cause damage to the delicate photovoltaic cells.

The time required for recharging will depend on the individual characteristics of the Ni-Cad cells being recharged, the current output of the photovoltaic cells, and the light intensity. Typically, about eight hours will be required for a full recharge.

Let me repeat the warning about the current, because it is very important. The current supplied by the photovoltaic cells must not exceed the maximum safe charging current specified by the manufacturer in the data sheet for the Ni-Cad cells you are attempting to charge.

NEVER attempt to recharge alkaline cells. They could explode if you attempt to recharge them. Even if you are lucky enough to avoid an explosion, this type of cell will not accept a recharge. DO NOT ATTEMPT IT! Trying to force a recharge on an alkaline battery could be very, very dangerous.

PROJECT 4: OVER-VOLTAGE INDICATOR

In many applications, we need to know when a voltage exceeds a specific value. For example, we might need to protect against an uncontrolled increase in the line power. In other applications, we are interested in a signal only when it is fairly large. Signals below the reference level can simply be ignored.

Of course, we could sit around and watch a voltmeter, but that would be tedious at best, and probably totally impractical in most applications. We need a circuit to automatically monitor the voltage and just alert us when it goes above the specific reference level for our particular application.

A circuit for this purpose is shown in Fig. 5-5. As you can see, this is a very simple project. The parts list is given in Table 5-4. Note that no wattage rating is given for

Fig. 5-5. Project 4: over-voltage indicator.

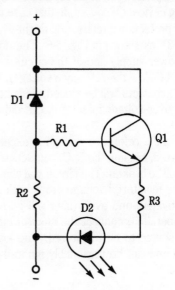

**Table 5-4. Parts List for the
Over-Voltage Indicator Project of Fig. 5-5.**

Component	Description
Q1	npn transistor (2N2222, or similar)
D1	zener diode to suit application
D2	LED
R1	470-ohm resistor
R2	4.7K resistor
R3	180-ohm, 1/4-watt resistor

two of the resistors (R1 and R2). The desired wattage depends on the specific application. In a low power system, $\frac{1}{4}$-watt resistors can be used. In a higher-powered system, you should use $\frac{1}{2}$-watt resistors. Of course, you could use $\frac{1}{2}$-watt resistors even in a low-powered system, although they are somewhat larger and will increase the physical size of the circuit slightly. This probably won't matter in most hobbyist applications.

Almost any npn transistor can be used as Q1. The 2N2222 specified in the parts list is a fairly low-powered device. In high-power applications, you should substitute a heftier transistor.

The only really critical component in this project is the zener diode (D1). This diode must be selected for your specific application. The breakdown voltage of the zener diode is the reference voltage for the over-voltage monitor circuit. When the breakdown voltage of the zener diode is exceeded, the LED (D2) will light up.

If you prefer, you could replace the LED with a piezoelectric buzzer or other audible signalling device. Of course, in that case this would no longer be an optoelectric project.

For protection circuits, you might want to place a relay in parallel with the indicating LED (D2), as shown in Fig. 5-6. The normally closed contacts of the relay are placed in the power-supply line of the circuit being protected. As long as the voltage is at a safe level (below the reference voltage), the relay will be in a deactivated condition, and the N.C. contacts will be closed, permitting the protected circuit to function normally, as if the over-voltage monitor circuit didn't exist at all. Of course, the LED will remain dark.

When the reference voltage is exceeded, the LED will light up, and the relay will be activated, causing the N.C. contacts to open up. The voltage supply to the protected circuit will be broken. The circuit will shut down. This system is automatically resetting. When the monitored voltage drops back down below the reference level, the relay will be deactivated, and everything will go back to normal.

In operation, connect the monitor circuit positive (+) and negative (−) across the voltage source to be monitored. The load this circuit places on the voltage source is minimal, and can be reasonably ignored in all but the most critical applications.

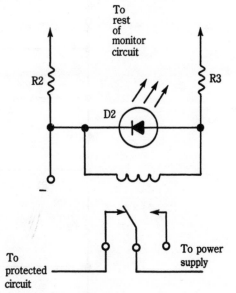

Fig. 5-6. For protection circuits, place a relay in parallel with the LED.

PROJECT 5: LOW-VOLTAGE ALERT

In a sense, this project is just the opposite of Project 4. This time the indicator LED is turned on when the monitor voltage drops <u>below</u> the reference voltage. A higher voltage will keep the LED turned off.

This project could be used to watch for a signal that might drop below a specific, critical level, or it can be used as a warning for battery-powered equipment. As batteries near the end of their useful life, the voltage will start to drop. When the warning LED lights up, the operator knows it is time to change the batteries before they die completely.

(This project isn't as useful when rechargeable Ni-Cad batteries are being monitored. These tend to put out the full voltage until they are very close to being fully discharged, and then the voltage drops off very rapidly. This project can still be used, but you'll have a much shorter warning period before the equipment stops working.)

The schematic for this low-voltage alert project is shown in Fig. 5-7. The parts list is given in Table 5-5. For more details, refer back to Project 4. The same information applies here. Of course, the reference voltage is set by the breakdown voltage of the zener diode (D1).

In operation, connect the monitor circuit positive (+) and negative (−) across the voltage source to be monitored. The load this circuit places on the voltage source is minimal, and can be reasonably ignored in all but the most critical applications.

PROJECT 6: HIGH/LOW-VOLTAGE MONITOR

In some applications, we need to know if a signal goes outside a specified range. As long as the monitored voltage is within this range of values, we can ignore it, but

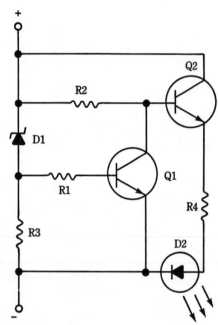

Fig. 5-7. Project 5: low-voltage alert.

**Table 5-5. Parts List for the
Low-Voltage-Alert Project of Fig. 5-7.**

Component	Description
Q1, Q2	npn transistor (2N2222, or similar)
D1	zener diode to suit application
D2	LED
R1	470-ohm resistor
R2, R3	4.7K resistor
R4	180-ohm, 1/4-watt resistor

we need to be alerted to take some action if the signal voltage goes outside the specified range (too low or too high).

You could use both Project 4 and Project 5 to monitor for too high and too low conditions, but that is a rather inelegant solution. This project takes an entirely different approach. In this circuit, which is illustrated in Fig. 5-8, we use a modified window comparator.

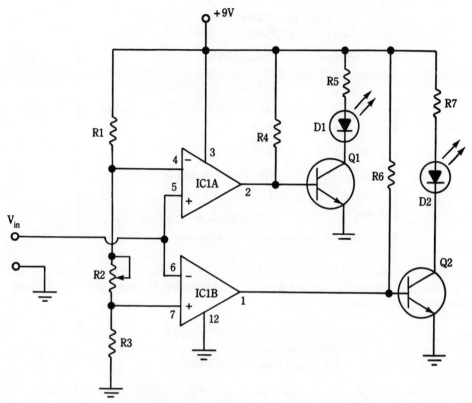

Fig. 5-8. Project 6: high/low-voltage monitor

A window comparator indicates when the input voltage is within a specified range. This circuit is modified to reverse the indication. When the signal voltage goes more positive than the upper end of the acceptable range (window), LED1 lights up. If the signal voltage goes more negative than the lower end of the acceptable range, LED2 lights up. If neither LED is lit, then the input voltage is somewhere within the acceptable range.

For maximum ease of reading the comparator outputs, it might be helpful to use two different colored LEDs for LED1 and LED2.

Potentiometer R2 sets the width of the acceptable range, or window. In some applications, this may be a front-panel control. In others, you might want this range-setting control to be a "set and forget" trimpot. You could even replace the potentiometer with a fixed resistor.

**Table 5-6. Parts List for the High/
Low-Voltage Monitor Project of Fig. 5-8.**

Component	Description
IC1	LM339 quad comparator
Q1, Q2	npn transistor (2N2222, or similar)
D1, D2	LED
R1, R3	27K, 1/4-watt resistor
R2	25K potentiometer, or trimpot
R4, R6	10K, 1/4-watt resistor
R5, R7	330-ohm, 1/4-watt resistor

You might also want to consider experimenting with the values of resistors R1 and R3. These two resistances interact with the potentiometer, but as a rough rule of thumb, they serve the following functions:

R1	set upper limit of window
R2	(potentiometer) set window width
R3	set lower limit of window

A typical parts list for this project appears in Table 5-6.

Chapter 6

Control Circuits

The projects in this chapter are designed to perform various switching and other control functions in electronics systems. These projects provide light-activated control over various electrical parameters, and can be used in a wide variety of practical applications.

PROJECT 7: LIGHT-OPERATED RELAY

Virtually any electrically powered device can be controlled with this project. The control in this case is in the form of switching. A relay is used to control the load circuit. The circuit for controlling the relay with light is shown in Fig. 6-1. A typical parts list is shown in Table 6-1.

A phototransistor (Q1) is used as the sensor. When the light striking the phototransistor exceeds a specific level, the relay is activated. When the light level drops below the trigger point, the relay is deactivated. The trigger point, or the sensitivity of the circuit, is set by potentiometer R1.

Transistor Q2 is a simple amplification stage to boost the output signal from the phototransistor (Q1) sufficiently to activate the relay.

Diode D1 is used to protect against back EMF (voltage) that can be generated across the coil of the relay.

For many applications, some sort of light shield might need to be mounted over the sensor (Q1) to reduce problems of false triggering from uncontrolled ambient light sources.

The relay's contacts should be suitable for the desired load. If you need to control a high-current load, you can use one relay to control a second, larger relay, as illustrated in Fig. 6-2.

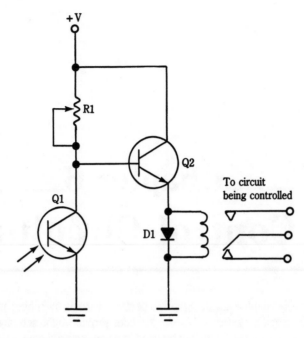

Fig. 6-1. Project 7: light-operated relay.

**Table 6-1. Parts List for the
Light-Operated Relay Project of Fig. 6-1.**

Component	Description
Q1	npn phototransistor
Q2	npn transistor (2N3904, or similar)
D1	diode (1N4002, or similar)
R1	100K potentiometer
K1	relay coil—500 ohm, 9 volt contacts—to suit load

PROJECT 8:   DARK-CONTROLLED RELAY

This project operates the opposite as the light-controlled relay of project 7. This time the relay is activated when the sensor is dark, rather than when it is illuminated.

Figure 6-3 shows the circuit for this project. The parts list is given in Table 6-2.

Note how similar this project is to the circuit shown back in Fig. 6-1. The only difference is the relative placement of the phototransistor sensor (Q1) and the sensitivity control (potentiometer R1).

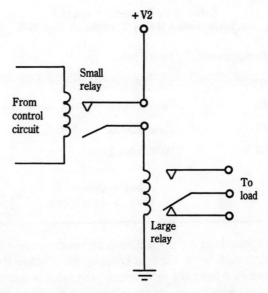

Fig. 6-2. A small relay can be used to drive a larger relay to handle heavier loads.

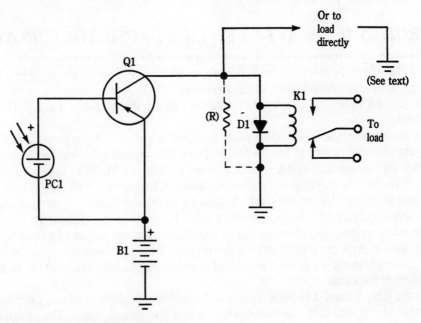

Fig. 6-3. Project 8: dark-controlled relay.

**Table 6-2. Parts List for the
Dark-Operated Relay Project of Fig. 6-3.**

Component	Description
Q1	npn phototransistor
Q2	npn transistor (2N3904, or similar)
D1	diode (1N4002, or similar)
R1	100K potentiometer
K1	relay coil—500 ohm, 9 volt contacts—to suit load

As long as the light shining on the sensor is above the trigger level (set by R1), the relay will be deactivated. When the light intensity drops below the trigger level, or when the sensor is shaded from the light source, the relay is activated.

As in project 7, the relay's contacts for this project should be selected to suit the desired load. If you need to control a high-current load, you can use one relay to control a second, larger relay, as illustrated in Fig. 6-2.

PROJECT 9:   HEAVY-DUTY, LIGHT-ACTIVATED RELAY

In some applications, we might want to control a larger load than projects 7 or 8 can handle. We could cascade a pair of relays, as discussed in the earlier projects. This project, however, offers a more elegant solution.

Figure 6-4 shows the circuit for a heavy-duty light-activated relay. Table 6-3 is the parts list.

A photovoltaic cell (PC1) serves as the sensor in this project. A pnp power transistor (Q1) amplifies the output of the photocell. The exact amount of power at the output depends on the specific transistor used, of course. Using the HEP230 device specified in the parts list, the amplifier can put out up to one full ampere, which should be more than enough to drive almost any relay a hobbyist is ever likely to use. If your load draws more than a quarter amp or so, you must use a heatsink on the transistor.

In many applications, the relay can be eliminated, and the output of transistor Q1 can be used to drive the load directly. If this is done, the load impedance should be very low, preferably under 30 ohms. Larger load impedances can interfere with the correct operation of the circuit.

If the relay is used, a resistor (shown in dotted lines) might be necessary in parallel with the relay coil to drop the load resistance to a low enough value. The resistance must be determined from the rated resistance of the relay coil. Remember the formula for two resistances in parallel:

$$1/R_t = 1/R1 + 1/R2$$

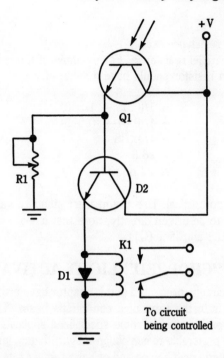

Fig. 6-4. Project 9: heavy-duty light-activated relay.

**Table 6-3. Parts List for the Heavy-Duty,
Light-Activated Relay Project of Fig. 6-4.**

Component	Description
Q1	pnp power transistor (HEP230, or similar)
D1	1N4002 diode
PC1	photovoltaic cell
B1	6-volt battery
K1	relay to suit load
(R)	optional parallel resistor

Algebraically rearranging this formula, we can solve for the necessary parallel resistance;

$$1/25 = 1/R + 1/KR$$
$$0.04 = 1/R + 1/KR$$
$$1/R = 0.04 - 1/KR$$
$$R = 1/(0.04 - 1/KR)$$

where
    KR is the relay coil resistance, and
    R is the necessary parallel resistance. For example, if the coil resistance is rated
at 500 ohms, the parallel resistor should have a value of:

$$
\begin{aligned}
R &= 1/(0.04 - 1/500) \\
  &= 1/(0.04 - 0.002) \\
  &= 1/0.038 \\
  &= 26.3 \\
  &\cong 27 \text{ ohms}
\end{aligned}
$$

The exact resistance is not critical. Use the nearest standard value.
    If the load device is to be driven directly, note that it must share the +6-volt dc
supply voltage used by the amplifier (Q1).

PROJECT 10: "CHOPPED" LIGHT-ACTIVATED RELAY

So far the switching circuits presented in this chapter have involved the assumption
that the activating light is in the form of a continuous beam. This is fine for many
applications, but can be problematic in some specialized applications.
    In some circuits, it is preferable to use *chopped* light, that is, instead of a continuous
beam, the light is modulated (switched on and off) at a regular rate. Generally this chop
rate is in the mid-audio range. Such systems typically use a frequency between about
400 Hz and 1 kHz (1000 Hz).
    A control circuit for a relay activated by a chopped light source is shown in Fig.
6-5. The parts list for this project appears in Table 6-4.

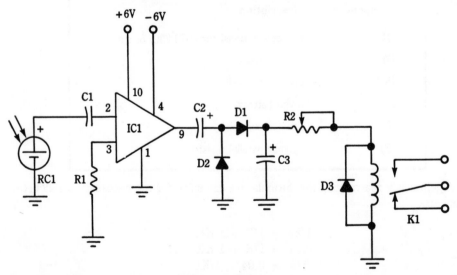

Fig. 6-5. Project 10: "chopped" light-activated relay.

Table 6-4. Parts List for the "Chopped"
Light-Activated Relay Project of Fig. 6-5.

Component	Description
IC1	op amp (CA3010, or similar)
D1, D2	diode (1N34A, or similar)
D3	diode (1N4002, or similar)
PC1	photovoltaic cell
C1	15 μF, 15-volt electrolytic capacitor
C2	1 μF, 15-volt electrolytic capacitor
C3	2 μF, 15-volt electrolytic capacitor
R1	2.7K, 1/4-watt resistor
R2	10K potentiometer (wire-wound type recommended)

The relay (K1) should be selected to suit the desired load circuit. Potentiometer R2 serves as a sensitivity control, determining the light intensity required to activate the relay.

Note that the op amp (IC1) requires a dual-polarity power supply (± 6 volts).

At low modulation or chop rates, there can be a slight problem with relay "chatter." If this type of problem shows up, you can improve the circuit's performance by increasing the value of capacitor C3 slightly.

PROJECT 11: LIGHT-CONTROLLED CAPACITANCE

Many electrical parameters can be controlled by a variable resistance. In such cases, we can use a photoresistor to place that parameter under light control (see Chapter 1). In some cases, however, the application might call for a variable capacitance instead of a variable resistance. There aren't any photocapacitors on the market (at least, not that I know of).

This project can be used in many such applications. Figure 6-6 shows the schematic diagram for this circuit. The parts list appears in Table 6-5.

A photovoltaic cell (PC1) is the light sensor. A capacitance that varies in response to the light intensity will appear across the output terminals.

The heart of this circuit is the varactor (D1). Many hobbyists are not very familiar with varactor diodes, so it might be helpful to describe this type of component briefly here.

A *varactor diode* produces an electrically variable capacitance. Of course, any diode exhibits some amount of capacitance when it is reverse-biased. In standard diode applications, this capacitance can be undesirable, because it can limit the operating

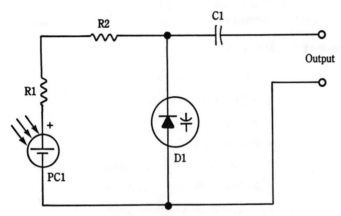

Fig. 6-6. Project 11: light-controlled capacitance.

**Table 6-5. Parts List for the
Light-Controlled Capacitance Project of Fig. 6-6.**

Component	Description
D1	varactor (1N4815A, or similar)
PC1	photovoltaic cell
C1	0.022-μF capacitor
R1	2.2-megohm, 1/4-watt resistor
R2	3.3-megohm, 1/4-watt resistor

frequency of the diode, especially in detector or rectifier applications. A varactor diode, on the other hand, turns this undesirable capacitance into a desired and controllable characteristic. Applications for varactor diodes include frequency modulation (FM) of oscillators, frequency multipliers, and multiwavelength tuners for rf amplifiers. Because the varactor diode behaves as an electrically controllable capacitance, this component is sometimes called a *varicap*.

A high-value isolating resistor is usually needed in varactor applications. This is the function of resistors R1 and R2. Two series resistors are used for convenience. They are easier to find than a single higher-value unit. In addition, higher-valued resistors can have some stability problems.

Capacitor C1 blocks any dc from the photocell from reaching the output terminals. It also makes it impossible for the external load circuit to place a potentially harmful short circuit across the varactor diode (D1). Capacitor C1's value is considerably higher than the maximum varactor capacitance, so it can effectively be ignored.

The range of output capacitances available depends on the specific photocell (PC1) and varactor diode (D1) used. The output capacitance is fairly low in value. Typically, the range of output capacitances runs from a little over 100 pF up to 250 pF, or so.

The light-controlled capacitance circuit can be used anywhere you might use a small-valued variable capacitor.

PROJECT 12: LIGHT-CONTROLLED MOTOR

With a little imagination and ingenuity, virtually any electrically powered device can be placed under light control. Figure 6-7 shows a circuit for a light controller for a small dc motor. The parts list is given in Table 6-6.

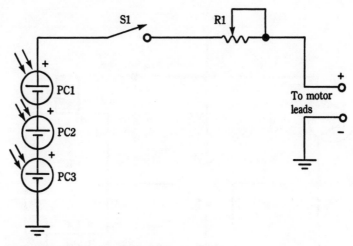

Fig. 6-7. Project 12: light-controlled motor.

**Table 6-6. Parts List for the
Light-Controlled Motor Project of Fig. 6-7.**

Component	Description
PC1, PC2, PC3	photovoltaic cell
R1	50-ohm wire-wound potentiometer
S1	SPST switch

The speed of the motor is controlled by the light intensity. If the light level is too low (dark), the motor will not run at all.

Potentiometer R1 is a wirewound type. It is used as a sensitivity control. When the light intensity is constant (simple on/off control), this potentiometer operates as a speed control. The potentiometer can be optional in some applications.

The motor should be a small, low-friction-bearing dc unit. Three photovoltaic cells are shown here for a supply voltage of approximately 1.5 volts. If the load motor requires a higher voltage, more photocells will be required. Obviously, there is a definite practical limit to the motor's size. Its voltage and current requirements must be minimal.

Switch S1 is just a simple power on/off switch, permitting you to turn off the motor regardless of the ambient light intensity. In some applications, this switch can be omitted.

PROJECT 13:   LIGHT-BEAM INTERRUPTION DETECTOR

The circuit shown in Fig. 6-8 can be used to detect any interruption in a light beam. Table 6-7 gives the parts list for this project.

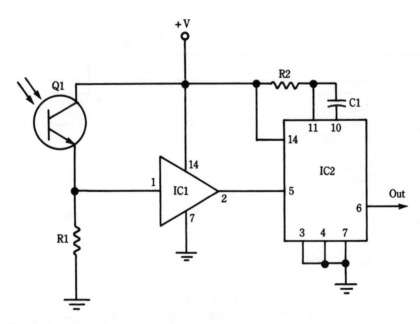

Fig. 6-8. Project 13: light-beam interruption detector.

In operation, the beam from the light source is focused on the phototransistor (Q1), as shown in Fig. 6-9A. When an object passes between the light source and the sensor, as shown in Fig. 6-9B, a shadow is cast upon the sensor. It will be dark, as long as the object is in the way. When this happens, a voltage equivalent to a logic HIGH is placed on the input of the Schmitt trigger (IC1), which, in turn, triggers the monostable multivibrator (IC2). The output then goes high for a specific timing period (set by resistor R2 and capacitor C1).

This circuit generates a fixed-length output pulse anytime the light beam is interrupted, no matter now briefly. This pulse can be used to trigger external circuitry, depending on the specific application.

**Table 6-7. Parts List for the Light-Beam
Interruption Detector Projects of Fig. 6-8.**

Component	Description
.IC1	74C14 Schmitt trigger
IC2	74C121 monostable multivibrator
Q1	npn phototransistor
C1	0.01-μF capacitor
R1	100 ohm, 1/4-watt resistor
R2	18K, 1/4-watt resistor

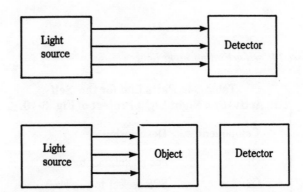

Fig. 6-9. If an object passes between the light source and the sensor, the beam is broken.

PROJECT 14: SELF-ACTIVATING NIGHT LIGHT

This project is used to illuminate automatically any area that might go dark. As long as there is sufficient ambient light, the night light will remain off. It will only go on when it is needed.

The schematic diagram for this project appears in Fig. 6-10. Table 6-8 shows the parts list.

This project is entirely straightforward. The phototransistor (Q1) serves as the light sensor. As long as it is sufficiently illuminated, the phototransistor shunts the SCR's (Q2) gate current to ground, and nothing happens.

When the ambient lighting level drops below a specific trigger point, however, the phototransistor sends current into the gate of the SCR (Q2), turning it on. When the SCR is turned on, it permits power to flow through the ac socket, and whatever load is plugged into it.

Potentiometer R2 sets the sensitivity of the detector. Adjust this control for the desired turn-on point.

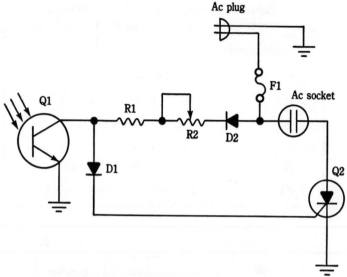

Fig. 6-10. Project 14: self-activating night light.

Table 6-8. Parts List for the Self-Activating Night Light Project of Fig. 6-10.

Component	Description
Q1	npn phototransistor
Q2	SCR (select to suit load)
D1, D2	diode (1N5059)
R1	1-megohm, 1/4-watt resistor
R2	5-megohm potentiometer
F1	fuse to suit load and SCR
ac socket	
ac plug	

The SCR should be selected to provide enough current to the desired load. It is best to over-rate the SCR slightly, that is, if you want to drive a load that draws 3.5 amps, use an SCR that can supply at least 4 amps, if not more.

Select the fuse to suit the desired load. For safety's sake, do not omit the fuse from this circuit. Do not use too large a fuse, or it won't serve any purpose.

Remember, this is an ac-powered circuit. Use all necessary precautions. Make 100% sure that no one can receive a shock from the circuit under any circumstances. No conductor anywhere in the circuit should be exposed.

PROJECT 15: LIGHT DIMMER

Light dimmers are always popular projects. They're great for adjusting the lighting in a room to suit the desired mood.

Figure 6-11 shows a simple and inexpensive light dimmer circuit. Table 6-9 shows the parts list for this project.

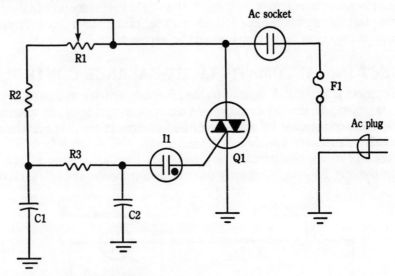

Fig. 6-11. Project 15: light dimmer.

Table 6-9. Parts List for the Light Dimmer Project of Fig. 6-11.

Component	Description
Q1	triac (40502, or select to suit load)
C1, C2	0.068-μF capacitor (250 volts, or better)
R1	50K potentiometer
R2	10K, 1/4-watt resistor
R3	15K, 1/4-watt resistor
I1	neon lamp (NE-2)
ac socket	
ac plug	

Potentiometer R1 is used to control the lighting level. The triac (Q1) is triggered via the neon lamp. In operation, adjust potentiometer R1 until the neon lamp conducts. This generally occurs when the lights are at about half-brilliance. Then you can back down on the light intensity until the lights are dimmed to the desired level.

The triac (Q1) is selected to provide sufficient power to the desired load. It is best to over-rate the triac somewhat. I strongly advise a heatsink for this component. A good triac to use in this application is the 40502, manufactured by RCA. With proper heatsinking, this triac can drive loads of up to 400 watts.

Select the fuse to suit the desired load. For safety's sake, do not omit the fuse from this circuit. Do not use too large a fuse, or it won't serve any purpose.

Remember, this is an ac-powered circuit. Use all the necessary precautions. Make 100% sure that no one can receive a shock from the circuit under any circumstance. No conductor anywhere in the circuit should be exposed.

PROJECT 16:   AUTOMATIC LIGHT-BALANCE CONTROLLER

This project is an unusual variation on the basic light dimmer presented as project 15. In this project, the dimmer circuit senses the ambient light level and automatically adjusts itself to compensate for any uncontrolled changes in the lighting. The circuit automatically balances the lighting to a desired level.

Figure 6-12 shows the circuit for this project. The parts list appears in Table 6-10. Potentiometer R1 is the sensitivity control, setting the desired lighting level.

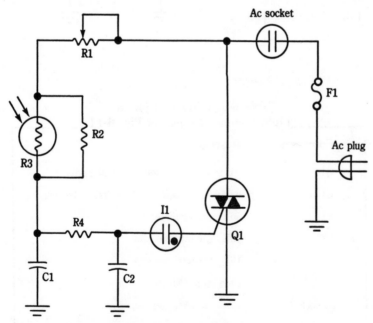

Fig. 6-12. Project 16: automatic light balance controller.

**Table 6-10. Parts List for the
Automatic Light Balance Controller Project of Fig. 6-12.**

Component	Description
Q1	triac (40502, or select to suit load)
C1, C2	0.068-μF capacitor (250 volts, or better)
R1	10K potentiometer
R2	470K, 1/4-watt resistor
R3	photoresistor
R4	15K, 1/4-watt resistor
I1	neon lamp (NE-2)
ac socket	
ac plug	

Overall, this circuit is the same as project 15, except for the addition of the photoresistor (R3). Refer to the text for project 15 for circuit details.

Remember, this is an ac-powered circuit. Us all necessary precautions. Make 100% sure that no one can receive a shock from the circuit under any circumstances. No conductor anywhere in the circuit should be exposed.

PROJECT 17: UNIVERSAL VOLTAGE CONTROL

Voltage control is a handy concept widely employed in such far-flung fields as automation and electronic music. It is easy to generate electrically a voltage of almost any desired pattern. This voltage is used to control some other electrical parameter in another circuit.

A simple form of voltage control can be added to almost any circuit with an optoisolator. This system is illustrated in its simplest form in Fig. 6-13. The optoisolator in this case is made up of a light source and a photoresistor. The signal source may be either an incandescent lamp or an LED. The voltage applied to the light source controls its brightness. The intensity of the light in turn controls the resistance of the photoresistor. The photoresistor can be used in place of a potentiometer, or even a fixed resistor in almost any circuit.

In this super-simple form, you will probably have to create your own optoisolator from a discrete light source and photoresistor. Optoisolators with simple photoresistors as the sensing/output device are not commonly available commercially. If you prefer to use a commercial optoisolator, use one with a phototransistor as the sensing/output device (see Fig. 6-14).

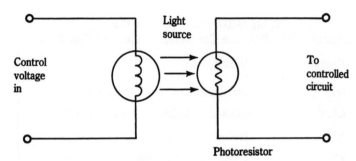

Fig. 6-13. Project 17: universal voltage control.

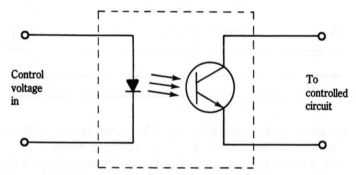

Fig. 6-14. An optoisolator can be used in the universal voltage-control project.

The amount of light shining on the phototransistor serves as the base signal, controlling the collector/emitter current. Thanks to the interrelationship of voltage, current, and resistance defined by Ohm's Law, the net effect is very similar to a light variable resistance.

PROJECT 18:   LIGHT-CONTROLLED AMPLIFIER

The most basic and most frequently used type of electronic circuit is the amplifier. Amplifiers are used in one form or other in virtually every electronic system.

In this project, we will build an amplifier whose gain is controllable by light intensity. That is, the brightness of the light determines the volume or amplitude of the output signal. In some cases, this could be used as a primitive form of remote control.

The circuit for the light-controlled amplifier project is shown in Fig. 6-15. The parts list appears in Table 6-11.

As you can see, there is nothing all that special or complex about this circuit. The project is simply an inverting op-amp amplifier. A photoresistor is used as the feedback resistor (R2).

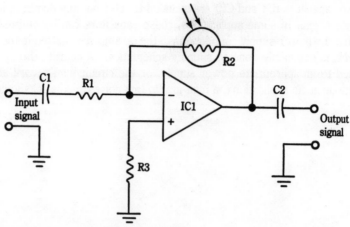

Fig. 6-15. Project 18: light-controlled amplifier.

Table 6-11. Parts List for the Light-Controlled Amplifier Project of Fig. 6-15.

Component	Description
IC1	op-amp IC
C1, C2	0.01-μF capacitor
R1	470K, 1/4-watt resistor
R2	photoresistor
R3	330K, 1/4-watt resistor

If you are up on op-amp theory, you should recall that the closed loop gain is determined by the ratio of the input and feedback resistances, according to this formula;

$$G = -R_f/R_i$$

where
G is the gain factor
R_f is the feedback resistance (R2), and
R_i (R1) is the input resistance

In this circuit, the feedback resistance varies in response to changes in the illumination of the photoresistor (R2). Therefore, the gain changes with each change in the feedback resistance, which changes with the light level. The gain of the amplifier is controlled by the amount of light directed at the sensor.

The two capacitors (C1 and C2) are included to filter out any dc offset in the input and/or output signal. In some applications, these capacitors can be omitted.

Note that the power-supply connections to the op amp are omitted in the schematic diagram. This is commonly done to simplify schematics. Of course, the op amp must be connected to an appropriate power supply, or the circuit won't work at all.

For serious audio applications, a high-grade, low-noise op-amp IC should be used.

Chapter 7

Sound and Audio Circuits

Most hobbyists enjoy building sound and tone-generator circuits. This chapter features several optoelectric projects with an audio component.

There is quite a bit of variety in the projects in this chapter. Included here are some of my personal favorites.

PROJECT 19: LIGHT-ACTIVATED TONE GENERATOR

The schematic for this project is shown in Fig. 7-1. Table 7-1 shows the parts list.

This circuit is built around the LM3909 IC. This chip will be used in several of the projects in Chapter 8, because it is intended primarily for LED flasher applications. We aren't really misusing the LM3909 here. It is officially designated as an ''LED flasher/oscillator'' IC, but we are certainly using it in a less common application.

The tone frequency is determined by capacitor C1, resistor R1, and potentiometer R2. Adjust the potentiometer for the desired pitch.

The LM3909 consumes very little current, so it is ideal for solar powered projects. Three photovoltaic cells (PC1 through PC2) are used as the power source. When the illumination level falls below a certain point, the speaker turns off. When the light strikes the photocells, (they should be mounted so that they are more or less equally illuminated), the speaker emits an audio tone. The light intensity might have a slight effect on the tone frequency in some cases, but the frequency in this circuit is controlled mainly by C1, R1, and R2, as discussed above.

Resistor R3 and potentiometer R4 control the output volume. These two resistances can be replaced by a single fixed resistor if you don't want to bother with a volume control. The higher the R3–R4 resistance is, the lower the volume will be. For practical applications, this resistance should be kept in the 45-ohm to 200-ohm range.

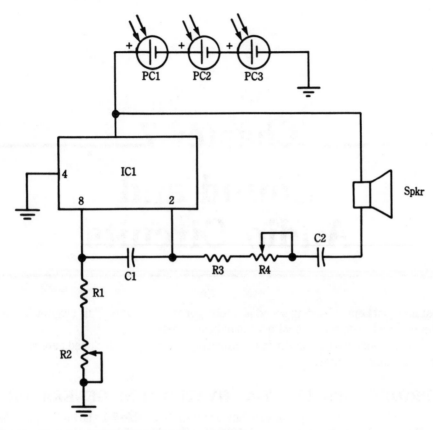

Fig. 7-1. Project 19: light-activated tone generator.

**Table 7-1. Parts List for the Light-
Activated Tone Generator Project of Fig. 7-1.**

Component	Description
IC1	LM3909 LED flasher/oscillator
PC1–PC3	photovoltaic cell
C1	0.1-μF capacitor
R1	6.8K, 1/4-watt resistor
R2	20K potentiometer
R3	47-ohm, 1/4-watt resistor
R4	100-ohm potentiometer
Spkr	small 8-ohm speaker

PROJECT 20: LIGHT-CONTROLLED TONE GENERATOR

In Project 19, the tone could be turned on and off by an external light source. This project goes one step further. The light intensity actually controls the frequency of the generated tone.

Figure 7-2 shows the circuit diagram for this project. A typical parts list for this project is given in Table 7-2, although you are encouraged to experiment with other component values.

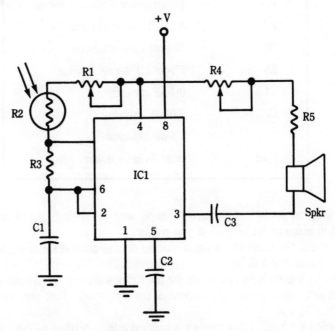

Fig. 7-2. Project 20: light-controlled tone generator.

There is nothing very fancy in this circuit. Basically, we just have a 555 timer (IC1) operating in the astable (rectangle-wave generator) mode. A photoresistor (R2) is inserted into the frequency-determining network (made up of R1 through R2, and capacitor C1). R2's resistance is dependent on the intensity of the light striking its surface. Therefore, controlling the light will control the frequency determining resistance. In other words, the output frequency is controlled by the intensity of the external light striking the photoresistor (R2).

Potentiometer R1 is added as a manual range control. This potentiometer can easily be replaced by a fixed resistor if you prefer. In some cases, you might be able to omit R1 from the circuit altogether.

Experiment with different values for capacitor C1. The value of this capacitor sets the overall range of the tone generator. You might use several capacitors, selected by a rotary switch to achieve manual selection of various ranges. The larger the value of capacitor C1, the lower the frequency range will be.

**Table 7-2. Parts List for the Light-
Controlled Tone Generator Project of Fig. 7-2.**

Component	Description
IC1	555 timer
R1	10K potentiometer
R2	photoresistor (cadmium sulfide)
R3	2.2K, 1/4-watt resistor
R4	250-ohm potentiometer
R5	82-ohm, 1/4-watt resistor
C1	0.1-μF capacitor
C2	0.01-μF capacitor
C3	0.5-μF capacitor
Spkr	small 8-ohm speaker

Potentiometer R4 is just a simple volume control. A single fixed resistor can be substituted in place of R4 and R5 if you prefer.

In operation, the brighter the light shining on the photoresistor is, the higher the output tone frequency will be.

Try shining a fixed light source on the photoresistor, and wave your hand over the sensor to create varying shadows to control the frequency. This project is a lot of fun to play with.

The basic idea behind this project is carried a little further in the solar Theremin project presented later in this chapter (Project 26).

Project 21: INFRARED-CONTROLLED OSCILLATOR

Like Project 20, this project is an oscillator whose output frequency is controlled by external light illuminating a photoresistor. The major difference in this case is that the photoresistor is not the ordinary cadmium-sulfide type. This photoresistor is made of lead-sulfide, which makes it sensitive to infrared light. This type of device is often marketed as an infrared detector.

The circuit is shown in Fig. 7-3, with the parts list appearing in Table 7-3.

Nothing is terribly critical here. Almost any matched npn and pnp transistors can be used. The output can drive an 8-ohm speaker directly, or it may be used as the input signal for some additional circuitry.

Potentiometer R2 allows the user to set the range of the output frequency. In some applications, this potentiometer can be replaced by a suitable fixed resistor or omitted altogether.

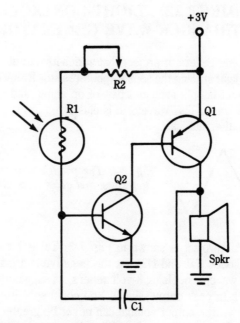

Fig. 7-3. Project 21: infrared-controlled oscillator.

**Table 7-3. Parts List for the
Infrared-Controlled Oscillator Project of Fig. 7-3.**

Component	Description
Q1	pnp transistor (2N3906, or similar)
Q2	npn transistor (2N3904, or similar)
C1	0.1-μF capacitor
R1	infrared sensitive photoresistor (lead sulfide—PbS)
R2	250K potentiometer

The output frequency varies with changes in the infrared light level. As the infrared light intensity increases, the output tone goes up in frequency.

Any object that gives off heat also emits infrared energy. You could consider this project to be a heat-controlled oscillator or an infrared-controlled oscillator. The two names ultimately mean pretty much the same thing.

Try aiming the sensor at a pot of boiling water, or a hot soldering iron. You should hear a very definite increase in the pitch of the tone produced by the speaker when the detector (R1) senses such hot, infrared-producing objects.

PROJECT 22: LIGHT-CONTROLLED TRIANGLE-WAVE GENERATOR

Here is another tone-generator project that varies its output frequency in response to the intensity of the light striking the sensor (photoresistor R5). The two big differences here are that the circuit is built around a pair of op amps, and the output signal is in the form of a triangle wave. This waveform is illustrated in Fig. 7-4. The tone quality is very mellow and rather flutelike.

Fig. 7-4. The triangle waveform has smooth, steep sides, and quick reversals of direction.

The circuit for this project is shown in Fig. 7-5. Table 7-4 shows the parts list.

The two zener diodes (D1 and D2) set the negative and maximum instantaneous values for the output waveform. Using 3-volt zeners, as suggested in the parts list, will result in a 6-volt peak-to-peak output signal. You could substitute other zener diodes if you prefer. Of course, the output voltage can never be greater than the supply voltage. If you use, say, 15-volt zeners, the output waveform will be severely clipped.

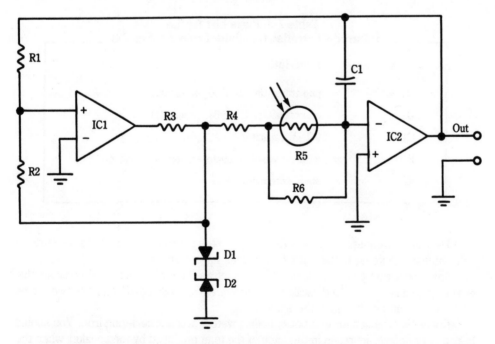

Fig. 7-5. Project 22: light-controlled triangle-wave generator.

Table 7-4. Parts List for the Light-Controlled Triangle-Wave Generator Project of Fig. 7-5.

Component	Description
IC1, IC2	op amp (101, or similar)
D1, D2	zener diodes (3 volt)
C1	0.1-μF capacitor
R1	8.2K, 1/4-watt resistor
R2	470K, 1/4-watt resistor
R3	10K, 1/4-watt resistor
R4	27K, 1/4-watt resistor
R5	photoresistor
R6	220K, 1/4-watt resistor

Resistor R6 is placed in parallel across the photoresistor (R5) to decrease its resistance to a range of values usable by the circuitry.

As with all audio projects, the best results will be obtained with a pair of high-grade, low-noise op-amp ICs.

You could try experimenting with the various parts values in the circuit. If resistor R4 is made variable (that is, if a potentiometer is used instead of a fixed resistor), it can be used as a manual range control for the output frequency. Changing the values of most of the other components could affect the purity of the output waveform. This might or might not be desirable, depending on your specific application.

PROJECT 23:   LIGHT-CONTROLLED FUNCTION GENERATOR

A function generator is an oscillator circuit that can produce two or more different output waveforms.

A schematic diagram for a simple but effective light-controlled function generator project is shown in Fig. 7-6.

Any of three basic waveforms can be selected via switch S1. These waveforms, which are illustrated in Fig. 7-7, are as follows:

1. Triangle wave
2. Square wave
3. Ascending sawtooth wave

A triangle wave is relatively weak in harmonics. A sawtooth wave is very harmonic-rich. A square wave is somewhere between these two extremes in terms of harmonic content.

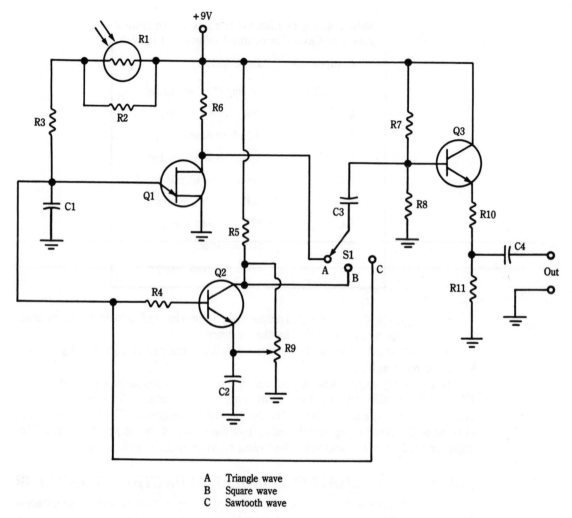

A   Triangle wave
B   Square wave
C   Sawtooth wave

Fig. 7-6. Project 23: light-controlled function generator.

The output frequency is controlled by the amount of light energy striking the photoresistor (R1). Resistor R1 is placed in parallel with the sensor to lower its overall effective resistance.

Potentiometer R9 controls the duty cycle of the square wave. This can be a calibrated trimpot, or a front panel control, depending on the requirements of your individual application.

The parts list for this project appears in Table 7-5. Experiment with other component values, especially for capacitor C1.

You might want to include several switch selectable capacitors in place of this single component to create manually selectable ranges for the output frequency.

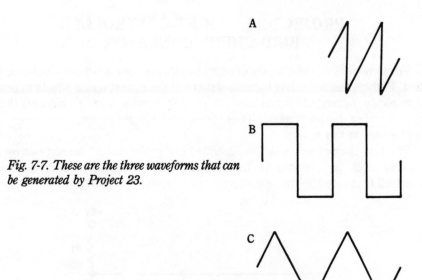

Fig. 7-7. These are the three waveforms that can
be generated by Project 23.

**Table 7-5. Parts List for the Light-
Controlled Function Generator Project of Fig. 7-6.**

Component	Description
Q1	UJT (unijunction transistor) (Radio Shack RS2029, or similar)
Q2, Q3	pnp transistor (2N4124, GE-20, or similar)
C1	0.05-μF capacitor
C2, C3	47-μF, 15-volt electrolytic capacitor
C4	0.47-μF capacitor
R1	photoresistor
R2	39K, 1/4-watt resistor
R3	1K, 1/4-watt resistor
R4, R5	10K, 1/4-watt resistor
R6	1.5K, 1/4-watt resistor
R7, R8	100K, 1/4-watt resistor
R9	50K potentiometer
R10, R11	2.2K, 1/4-watt resistor
S1	3 position, single-pole switch

PROJECT 24:   LIGHT-CONTROLLED
BIRD-CHIRP SIMULATOR

For some reason, the chirping of a bird has always been a popular electronic sound effect. Perhaps this is partially because it is one of the easiest sound effects to generate electronically. Around Christmas time you're likely to come across ornaments that produce a warbling, bird-like sound. These ornaments contain circuits that are not unlike the one used in this project.

This bird-chirp circuit, which is illustrated in Fig. 7-8, is built around two oscillator ICs. One oscillator generates the tone, while the second oscillator turns the first one on and off to create the chirping effect.

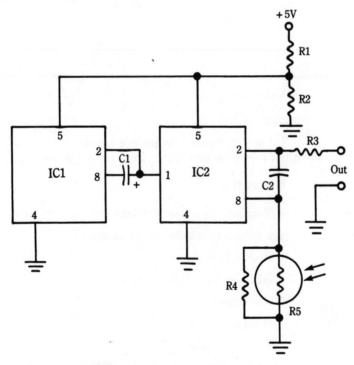

Fig. 7-8. Project 24: light-controlled bird-chirp simulator.

A photoresistor (R3) is used to control the tone frequency, that is, the pitch of the bird-like chirps is determined by the amount of light striking this sensor. The effective resistance of the photoresistor is reduced to a range of values usable by the circuit by the parallel resistor (R4).

The parts list for this project is given in Table 7-6.

Feel free to experiment with various component values. Capacitor C2, along with photoresistor R5, controls the tone frequency. Capacitor C1 sets the chirp rate. The smaller the value of this capacitor, the faster the chirps will be. If C1 is made small enough,

Table 7-6. Parts List for the Light-Controlled Bird-Chirp Simulator Project of Fig. 7-8.

Component	Description
IC1, IC2	3909 LED flasher/oscillator
C1 capacitor	33-μF, 15-volt electrolytic
C2	0.1-μF capacitor
R1	3.3K, 1/4-watt resistor
R2	1.5K, 1/4-watt resistor
R3	47-ohm, 1/4-watt resistor
R4	27K, 1/4-watt resistor
R5	photoresistor

frequency-modulation (FM) effects will begin to show up in the output signal. The sound will no longer be very bird-like, but you might find it interesting and useful for other purposes.

The output may be fed into a small audio amplifier, or it can be used to drive a small speaker directly.

PROJECT 25: "PUTT-PUTT" EFFECT GENERATOR

This circuit, shown in Fig. 7-9, can generate a number of sputtering effects, ranging from a slow "putt-putt-putt" effect to the roar of a racing car. The specific effect produced depends upon the amount of light energy striking the sensor (photoresistor R1).

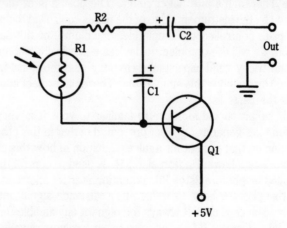

Fig. 7-9. Project 25: "putt-putt" effect generator.

**Table 7-7. Parts List for the
"Putt-Putt" Effect Generator Project of Fig. 7-9.**

Component	Description
Q1	pnp transistor (2N3906, or similar)
C1, C2	1-μF, 12-volt electrolytic capacitor
R1	photoresistor
R2	15K, 1/4-watt resistor

Resistor R2 prevents any possibility of damage to the transistor from too low a resistance.

Table 7-7 shows the parts list for this unusual project. Nothing is at all critical in this project. Almost any pnp transistor can be used for Q1.

Capacitors C1 and C2 also have a strong influence on the sound produced by this circuit. By all means, experiment with other values for these components. These capacitor values can be anywhere from 0.01 μF up to about 5 μF. The two capacitors don't necessarily need to have identical values. Many intriquing effects can be achieved by using mismatched capacitors.

PROJECT 26: SOLAR THEREMIN

This project is an updated and simplified variation of an early electronic musical instrument. The Theremin was a curious device that could be played without touching it. The original Theremins depended on hand capacitance. Two antennas or plates were mounted on the instrument. The player waved his hands over these two antennas or plates. The distance from the player's hand to the sensing surface created a capacitance that detuned the Theremin's internal circuitry. The positions of the hands controlled the sound produced by the instrument. One of the plates controlled the pitch or frequency, while the other plate controlled the amplitude, or volume, of the output signal.

In our version, we will use optoelectric devices (specifically, photoresistors) in place of the delicate and tricky hand-capacitance circuitry of the original instruments.

The schematic diagram for this simple solar Theremin project appears in Fig. 7-10. Table 7-8 shows the parts list.

Actually, this circuit should look pretty familiar to you. It is very, very similar to the light-controlled tone-generator project presented earlier in this chapter (Project 20). Refer to the section on that project for a full explanation of how this circuit works. The control circuits work as follows. Photoresistor R2 is used to control the frequency. The volume is controlled by photoresistor R6. Potentiometer R4 and resistor R5 are placed in parallel with this photoresistor to reduce its resistance significantly. By itself, the photoresistor's resistance is almost always too high for any audible tone to be produced by the speaker. Potentiometer R4 permits you to set a volume range for the instrument, which is a lot handier than you might suspect if you don't have any prior experience

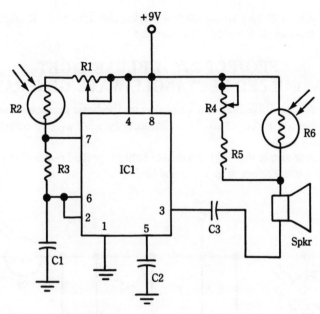

Fig. 7-10. Project 26: Solar Theremin.

**Table 7-8. Parts List for the
Solar Theremin Project of Fig. 7-10.**

Component	Description
IC1	555 timer
R1	10K potentiometer
R2, R6	photoresistor (cadmium sulfide)
R3	2.2K, 1/4-watt resistor
R4	1K potentiometer
R5	1K, 1/4-watt resistor
C1	0.1-μF capacitor
C2	0.01-μF capacitor
C3	0.5-μF capacitor
Spkr	small 8-ohm speaker

with a Theremin. If you prefer to omit this control, however, you can use a single fixed resistor in place of the R4–R5 combination in parallel with photoresistor R6.

A Theremin is a lot of fun to play. You can learn to play simple tunes very quickly. Position your hand over the R6 sensor to cut the volume to an inaudible level between

notes. While it is very easy to get started playing the Theremin, it takes a lot of work
to get really good at it. Practice, and have fun.

PROJECT 27:   DELUXE LIGHT-
CONTROLLED RECTANGLE-WAVE GENERATOR

This project and Project 28 are variations on the basic principles used in the solar
Theremin presented in Project 26. Two photoresistors are used to control the different
parameters of a sound.

The circuit shown in Fig. 7-11 is a deluxe light-controlled rectangle-wave generator.
A parts list for this project is given in Table 7-9.

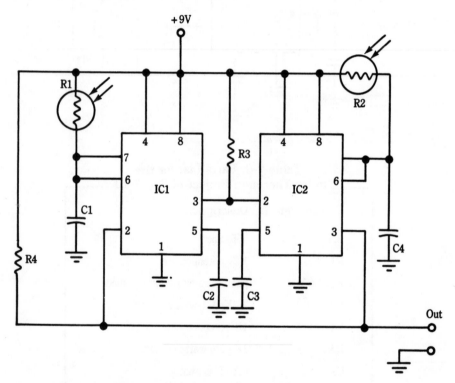

Fig. 7-11. Project 27: deluxe light-controlled rectangle-wave generator.

This project is made up of a pair of 555 timers. A 556 dual timer IC can be substituted,
if you prefer. Just be careful to correct the pin numbers.

IC1 and its associated components comprise a simple astable multivibrator, or
rectangle-wave generator. Photoresistor R1 controls the frequency, or the pitch. This
part of the circuit is basically the same as in Project 20.

The second timer (IC2) and its associated components control the duty cycle of the
rectangle wave. The duty cycle of a rectangle wave is a simple measure of how much

**Table 7-9. Parts List for the Deluxe Light-
Controlled Rectangle-Wave Generator Project of Fig. 7-11.**

Component	Description
IC1, IC2	555 timer
C1, C4	0.1-μF capacitor
C2, C3	0.01-μF capacitor
R1, R2	photoresistor
R3, R4	1.8K, 1/4-watt resistor

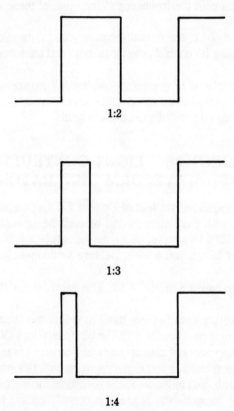

1:2

1:3

1:4

Fig. 7-12. Rectangle waves can have different duty cycles.

of each complete cycle is in the high state. Figure 7-12 shows some rectangle waves
with varying duty cycles.

The duty cycle defines the harmonic content of the signal. For simplicity, we will
consider only whole-number duty-cycle ratios.

A square wave is a rectangle wave with a duty cycle of 1:2. The signal is high for one half of each complete cycle. The signal contains only the fundamental and all odd harmonics. All harmonics that are multiples of two (even harmonics) are omitted.

A rectangle wave with a duty cycle of 1:3 is high for one third of each complete cycle. The signal contains the fundamental and all harmonics, except those that are multiples of three.

A rectangle wave with a duty cycle of 1:4 is high for one quarter of each complete cycle. The signal contains the fundamental and all harmonics, except those that are multiples of four.

This pattern holds true for any duty cycle. A duty cycle of 1:x means that harmonics that are whole number multiples of x are omitted.

In a simple 555 astable multivibrator circuit (like Project 20), the duty cycle can be changed, but it interacts with the frequency. When one of these parameters changes, so does the other.

In the circuit of Fig. 7-11, the output frequency and duty cycle are independently controllable. Photoresistor R1 controls the frequency, and photoresistor R2 controls the duty cycle.

This device can be played in the same "no hands" manner as the Theremin described in Project 26. Instead of amplitude, the second sensor is used to control the tonal quality or harmonic content of the output signal.

PROJECT 28:   LIGHT-CONTROLLED ODD-WAVEFORM GENERATOR

This project is an expanded version of Project 27. Once again, we have two 555 timers (IC1 and IC2), with the timing period of each being controlled by a separate photoresistor (R1 and R3). Of course, you may substitute a 556 dual timer IC for the two separate 555 timer ICs, if you prefer. Be sure to double-check the conversion of the pin numbers.

The circuit diagram appears in Fig. 7-13. The parts list for this project is given in Table 7-10.

Two six-position rotary switches are used to select the timing capacitor for each timer section. These controls, coupled with the photoresistors (R1 and R3) permit the operator to create a wide variety of unusual sounds. The effect is pretty hard to describe, but I strongly recommend building and experimenting with this project. It is fascinating and a lot of fun to play with. You might want to experiment further by substituting different values for some of the components. Nothing is terribly critical here.

This device can be played in the same "no hands" manner as the Theremin described in Project 26, but the effect in this case is really strange. This project almost makes a Theremin seem normal.

If you're really adventurous, you might go even further, and feed an external control voltage into pin #5 of one or both of the timer ICs. If you do this, eliminate the appropriate stabilizing capacitor (C7 or C8).

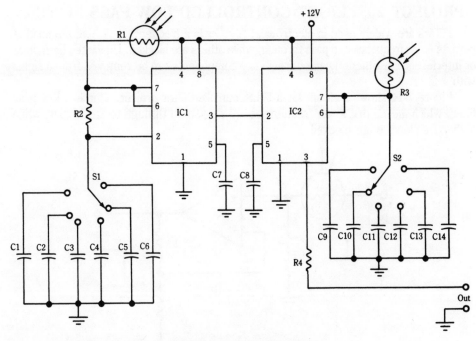

Fig. 7-13. Project 28: light-controlled odd waveform generator.

Table 7-10. Parts List for the Light-Controlled Odd-Waveform-Generator Project of Fig. 7-13.

Component	Description
IC1, IC2	555 timer
C1, C7, C8, C9	0.01-μF capacitor
C2, C10	0.047-μF capacitor
C3, C11	0.1-μF capacitor
C4, C12	0.47-μF capacitor
C5, C13	1-μF, 25-volt electrolytic capacitor
C6, C14	5-μF, 25-volt electrolytic capacitor
R1, R3	photoresistor
R2, R4	1K, 1/4-watt resistor
S1, S2	6 position, single-pole rotary switch

PROJECT 29:   LIGHT-CONTROLLED LOW-PASS FILTER

Filters are widely used in electronics. A *filter* is a frequency sensitive circuit that permits some frequencies to pass through, while others are blocked. Unwanted harmonics or interference or other unwanted frequency components can be removed from a signal with a filter.

Figure 7-14 shows a circuit for a light-controlled filter project. This is a low-pass filter, which means that lower frequencies are passed on through to the output, while higher frequencies are blocked.

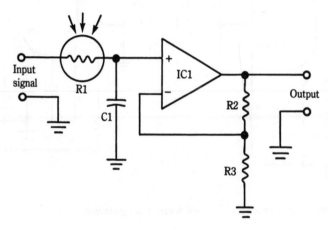

Fig. 7-14. Project 29: light-controlled low-pass filter.

Figure 7-15 shows a frequency-response graph for an ideal low-pass filter. A certain critical frequency is labeled F_c. This is the cutoff frequency of the filter. Anything below F_c is passed, while anything above F_c is blocked.

The graph of Fig. 7-15 is for an idealized filter. No practical filter approaches this ideal. In practical filters, there is a sloping intermediate range between the fully passed and fully blocked portions of the spectrum. A frequency-response graph for a more realistic

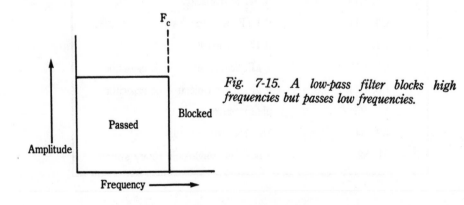

Fig. 7-15. A low-pass filter blocks high frequencies but passes low frequencies.

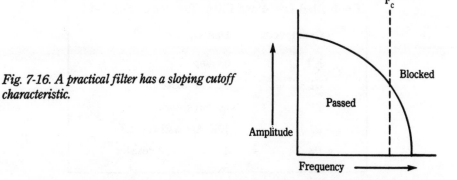

Fig. 7-16. A practical filter has a sloping cutoff characteristic.

low-pass filter is shown in Fig. 7-16. Note the intermediate slope. The steeper the slope is, the better the filter.

The circuit of Fig. 7-14 consists of two main parts—a passive filter stage and an amplifier stage. The amplifier compensates for losses in the filter. All frequencies passing through a passive filter network are attenuated to some extent. In many applications, some additional boosting of the passed portion of the signal is also desirable.

The filter portion of this circuit is made up of photoresistor R1 and capacitor C1. The basic, passive low-pass filter (with an ordinary fixed resistor instead of a photoresistor) is illustrated separately in Fig. 7-17. This simple filter has a fairly shallow slope.

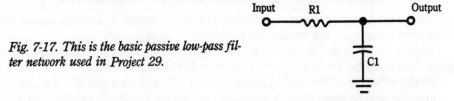

Fig. 7-17. This is the basic passive low-pass filter network used in Project 29.

The resistor and capacitor values determine the cutoff frequency (F_c) according to this simple formula:

$$F_c = 1/(2\pi R1C1)$$

π is a mathematical constant with a value of approximately 3.14. The formula can be rewritten as:

$$F_c = 1/(6.28R1C1)$$

Assuming the capacitance value (C1) doesn't change, any variation in the resistance value (R1) will cause the cutoff frequency (F_c) to change. The higher the resistance, the lower the cutoff frequency. In this project R1 is a photoresistor. This means the resistance varies with the light intensity. The amount of light striking the photoresistor controls the cutoff frequency of the filter.

**Table 7-11. Parts List for the Light-
Controlled Low-Pass Filter Project of Fig. 7-14.**

Component	Description
IC1	op amp
C1	0.22-μF capacitor
R1	photoresistor
R2	10K, 1/4-watt resistor
R3	4.7K, 1/4-watt resistor

The rest of the circuit (IC1, R2, and R3) is a fairly straightforward noninventing amplifier circuit. Resistors R2 and R3 set the gain of the circuit.

A typical parts list for this project appears in Table 7-11. You can experiment with other component values.

Feed a harmonic-rich signal into the input of this filter circuit, and notice how varying the light intensity striking the sensor (photoresistor R1) changes the tonal quality of the signal at the output. The light intensity controls the harmonic content of the output signal.

PROJECT 30:   DIGITAL SOLAR THEREMIN

This project is another light-controlled Theremin, similar in concept to the one featured in Project 26, but using a completely different approach.

As you can see in the circuit diagram of Fig. 7-18, this version is built around digital gates. The parts list for this project is given in Table 7-12.

In the earlier solar Theremin project (Project 26), it is tricky to get any silences (rests, or pauses) between notes. You can only reduce the amplifier's amplitude as much as possible, and hope for the best.

In this version, one of the photoresistors (PC1) functions as an on/off control. The performer can determine when a note will be sounded and when the instrument will remain silent. The trade-off here is that there is no control over the dynamics (moment-to-moment volume level) of the music. The second photoresistor (PC2) controls the frequency of the tone produced by the Theremin.

There is nothing terribly critical or complex about this circuit. The first portion of the circuit (including PC1, IC1A, IC1B, and IC1C) is a light-activated gate. When the amount of light striking the surface of the photoresistor (PC1) exceeds a specific level (preset by potentiometer R1), the output of IC1C goes low; otherwise, this output is high. A tone will be sounded only when PC1 is covered (in shadow from the performer's hand). When light is striking the photosensor, the Theremin remains silent.

The second half of the circuit (including IC1D, IC2A and their associated components) is a simple, gated digital oscillator circuit. Photoresistor PC2 controls the oscillator frequency. Therefore, the pitch of the tone from the speaker (heard only when

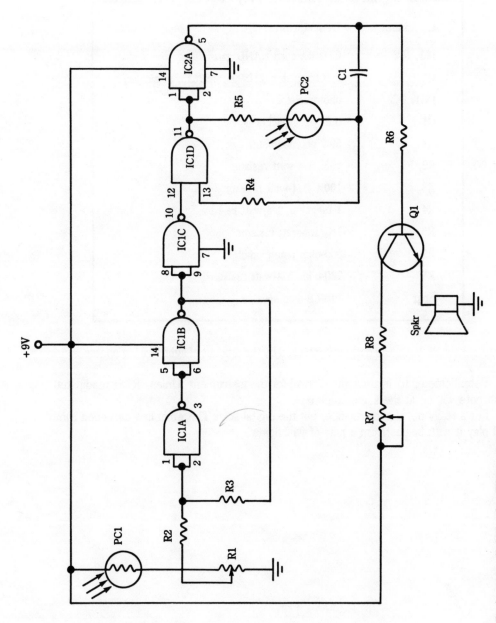

Fig. 7-18. Project 30: digital solar Theremin.

**Table 7-12. Parts List for the
Digital Solar Theremin Project of Fig. 7-18.**

Component	Description
IC1, IC2	CD4011 quad NAND gate
Q1	npn transistor (2N3904, 2N2222, or similar)
PC1, PC2	photoresistor
C1	0.01-μF capacitor
R1	50K potentiometer
R2, R6	10K, 1/4-watt resistor
R3	100K, 1/4-watt resistor
R4	1 megohm, 1/4-watt resistor
R5	1K, 1/4-watt resistor
R7	500-ohm potentiometer
R8	220-ohm, 1/4-watt resistor
Spkr	small 8-ohm speaker

photoresistor PC1 is in the dark) is controlled by the amount of light striking photoresistor PC2.

Potentiometer R7 is a volume control for the instrument. Unless R7 is readjusted, each note will be at the same amplitude.

For a really novel performance, put the digital solar Theremin in a darkened room and play it with beams from a pair of flashlights.

Chapter 8
LED Flashers

A popular type of the project is the LED flasher. There are countless variations on this basic, popular circuit.

Generally, the applications for LED flashers are rather lighthearted. They are used as novelty items and 'do-nothing' boxes. They can also be used as decorations and eye-catching displays, and in more serious applications, such as warning indicators.

PROJECT 31: BASIC LED FLASHER

Figure 8-1 shows a simple and inexpensive LED flasher circuit. Basically, it is just a low-frequency square-wave oscillator. The LED monitors the oscillator output. When the signal level goes high, the LED lights up. When the signal level goes low, the LED turns off. The circuit operation is certainly simple enough.

The oscillator frequency must be kept very low, generally under 10 Hz or so. At higher frequencies, you might not be able to distinguish between the individual flashes of the LED as it rapidly turns on and off, and it will appear to be continuously lit (possibly at a slightly lower than normal intensity).

Table 8-1 shows a typical parts list for this project. Experiment with other component values, especially for resistor R2 and capacitor C1. These are the primary frequency-determining components. Resistor R1 should have a value that is roughly ten times that of the series combination of R2 and R3. R1s' exact value is not very critical.

With the component values specified in the parts list, the flash rate is about one or two flashes per second, depending on the setting of potentiometer R3.

Resistor R4 is a current-limiting resistor to protect the LED (D1). Decreasing the value of R4 causes the LED to glow more brightly when it is lit. Of course, increasing the value of R4 reduces the LED's brightness. To prevent damage to the LED, R4's value should not be lower than about 100 ohms. Increasing R4's value beyond about 1K or so usually causes the LED to glow too dimly to be of much use.

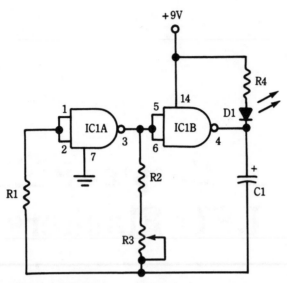

Fig. 8-1. Project 31: basic LED flasher.

Table 8-1. Parts List for the Basic LED Flasher Project of Fig. 8-1.

Component	Description
IC1	CD4011 quad NAND gate
D1	LED
C1	5-μF, 25-volt electrolytic capacitor
R1	1-megohm, 1/4-watt resistor
R2	10K, 1/4-watt resistor
R3	100K potentiometer
R4	470-ohm, 1/4-watt resistor

A simple variation on the basic LED flasher project appears in Fig. 8-2. This circuit is a gated LED flasher. An external logic signal can turn the oscillator on and off, controlling when the LED will blink, and when it will just sit there.

PROJECT 32: DUAL-LED FLASHER

A single flashing LED is certainly eye-catching, but two LEDs alternately flashing on and off would be almost impossible to ignore. A circuit for a dual-LED flasher project is shown in Fig. 8-3. A typical parts list for this project appears in Table 8-2.

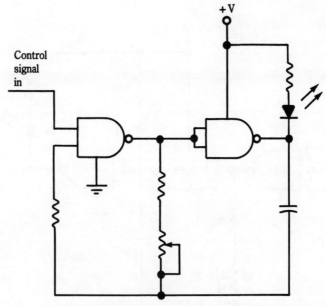

Fig. 8-2. A simple variation on Project 31 creates a gated LED flasher.

**Table 8-2. Parts List for the
Dual-LED Flasher Project of Fig. 8-3.**

Component	Description
IC1	CD4011 quad NAND gate
D1, D2	LED
C1	5-μF, 25-volt electrolytic capacitor
R1	1-megohm, 1/4-watt resistor
R2	10K, 1/4-watt resistor
R3	100K potentiometer
R4, R5	470-ohm, 1/4-watt resistor

Basically, this project is the same as project 31 with the addition of an extra inverter stage and a second LED (with its own current-limiting resistor, of course). Since the IC used is a quad gate and the original LED flasher project used just two sections, the extra inverter stage is already available. The only addition to the cost and bulk of the project is the second LED and a single resistor.

In operation, when LED1 is lit, LED2 will be dark, and vice versa. The LEDs will alternately flash on and off as long as power is applied to the circuit.

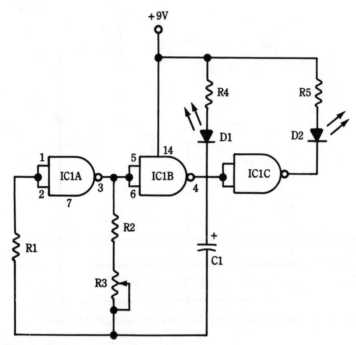

Fig. 8-3. Project 32: dual LED flasher.

PROJECT 33: EMERGENCY HIGH-INTENSITY FLASHER

A flashing LED is certainly eye-catching, but in many situations, it just isn't sufficient. There is a practical limit to the brightness of any LED's glow. Moreover, LEDs are rather small. If you need an indicator that must be visible from a distance, or under less than ideal viewing conditions, this project could be exactly what you need. It is a high-intensity flasher. It is particularly well-suited as an emergency-warning indicator.

The circuit diagram for this project appears in Fig. 8-4. Table 8-3 shows the parts list.

Do not be confused by the component labeled IC1. The symbol might look like an ordinary LED at a glance, but this is actually a special flasher LED. A miniature integrated-circuit oscillator is built into the LED's housing (see the section on flasher LEDs in Chapter 3).

When power is applied to the flasher LED (IC1), it starts to blink at a regular rate of approximately 3 Hz. In addition, the on/off pulses also appear across any component in series with the flasher LED. D1, which is an ordinary LED, blinks on and off in step with the flasher LED.

This LED (D1) might not appear to serve much purpose in this circuit. Actually, it is included to limit the voltage to the flasher LED (IC1). The LED (D1) could probably be replaced with an ordinary diode.

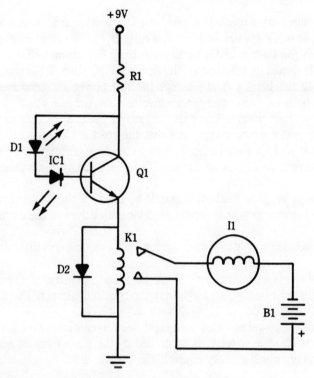

Fig. 8-4. Project 33: emergency high-intensity flasher.

Table 8-3. Parts List for the Emergency High-Intensity Flasher Project of Fig. 8-4.

Component	Description
IC1	flasher LED
Q1	npn transistor (2N2222, 2N3904, or similar)
D1	LED
D2	1N4002 diode, or similar
R1	150-ohm, 1/4-watt resistor
K1	dc relay coil—9-volt dc—500-ohm contacts—SPST—rated to suit load

The flasher pulses are amplified by transistor Q1, which can be almost any npn-type device. The output of the transistor controls a relay (K1). The relay contacts open and close in step with the flasher LED, turning the lamp (I1) on and off.

The lamp (I1) should be selected for the desired application. B1 is simply a suitable power source for the lamp you are using. Battery power is recommended for an emergency light, because many emergencies involve a loss (or unavailability) of ac power. For example, you might want to carry this project in your car in case of an accident. The flashing light will warn oncoming cars that the road is blocked.

If you must use an ac-powered lamp, be sure to use an appropriate relay with contacts rated for high-current ac use. A dc-powered lamp is strongly recommended for this project.

Separate power supplies (batteries) should be used for the flasher circuit and the output lamp. If a common power source is used for both halves of the circuit, reliability can suffer.

Diode D2 is included simply to protect against back-emf (voltage) that can form across the relay coil.

Of course, there is a practical limit to the size (and intensity) of the output lamp. The relay contacts must be able to safely carry the required current. The flasher circuit will not put out sufficient power to drive very large relays.

To drive a higher-powered lamp, you might need to cascade two relays, as shown in Fig. 8-5. The first, smaller relay, is controlled by the flasher circuit and drives the second, larger relay, which actually controls the output lamp.

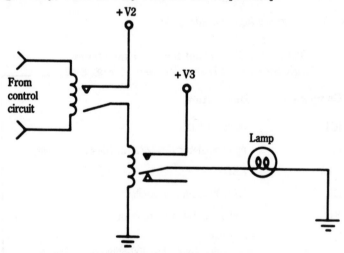

Fig. 8-5. A pair of relays can be cascaded to drive a larger lamp.

PROJECT 34: SEQUENTIAL COUNTDOWN FLASHER

The XR2240 programmable timer IC is a useful device for creating sequential flasher circuits. This chip is manufactured by Exar and Intersil. Basically, it is a timer followed by an eight-stage binary counter. The pinouts of this chip are shown in Fig. 8-6.

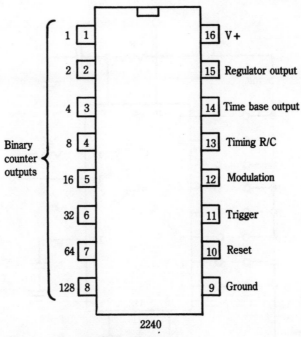

Fig. 8-6. This project is a programmable timer IC.

Pins #1 through #8 are the binary counter outputs. Each divides the basic time base value (determined by an external resistor and capacitor) by a factor of two. That is, in the monostable mode

$$T = RC$$

and the different pins have the following values:

Pin#	Time Value
1	1T
2	2T
3	4T
4	8T
5	16T
6	32T
7	64T
8	128T

The basic monostable circuit using the XR2240 is shown in Fig. 8-7. Only three external resistors and two external capacitors are required. Three of these components have pretty much standardized values, and are indicated in the diagram. The remaining

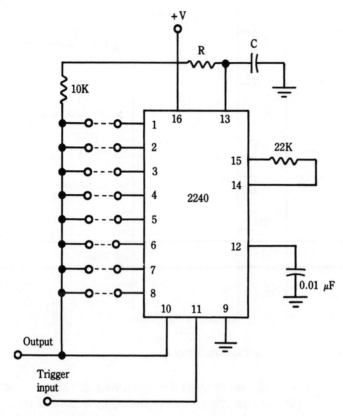

Fig. 8-7. This is the XR2240's basic monostable mode circuit.

two external components are the timing resistor and the timing capacitor, which determine the time base of the circuit.

For example, if R is a 220K resistor, and C is a 0.5-μF capacitor, the nominal time base will be:

$$T = RC$$
$$T = 220000 \times 0.0000005$$
$$= 0.11 \text{ second}$$

For an ordinary timer, such as the 555, that would be it. However, this time base is multiplied at each of the XR2240's outputs as follows:

Pin#	Time Multiplier Value	Output Time
1	1T	0.11 second
2	2T	0.22 second

3	4T	0.44 second
4	8T	0.88 second
5	16T	1.76 seconds
6	32T	3.52 seconds
7	64T	7.04 seconds
8	128T	14.08 seconds

Any of these outputs can be tapped off individually.

If you need an intermediate value, two or more of the counter outputs can be tied together. For example, by connecting pins #1, #3, and #6 together, the output time will be equal to:

$$1T + 4T + 32T = 37T$$

With a time base of 0.11 second, as in our example above, this works out to a timing period of 4.07 seconds.

For this sequential countdown flasher circuit, all eight of the XR2240's outputs are used independently. Each output pin drives its own LED. Figure 8-8 shows the complete circuit. The parts list for this project is given in Table 8-4. Experiment with other values for R1 and C1 to achieve different timing values. You might want to use a potentiometer for R1 to make the circuit manually adjustable.

When the timer is triggered by briefly closing switch S1, all eight output LEDs will light up. They will then go out one by one as their time periods expire. Using the component values given in the parts list, the time base is approximately:

$$T = RC$$
$$T = 680000 \times 0.00005$$
$$= 34 \text{ seconds}$$

The time each LED will remain lit is as follows:

LED#	Time	
1	34 seconds	0.56 minute
2	68 seconds	1.13 minutes
3	136 seconds	2.26 minutes
4	272 seconds	4.53 minutes
5	544 seconds	9.06 minutes
6	1088 seconds	18.13 minutes
7	2176 seconds	36.26 minutes
8	4352 seconds	72.53 minutes

It takes just under an hour and a quarter for the circuit to time itself out. Once the circuit has completely timed out, all of the LEDs will be dark again, until the timer is

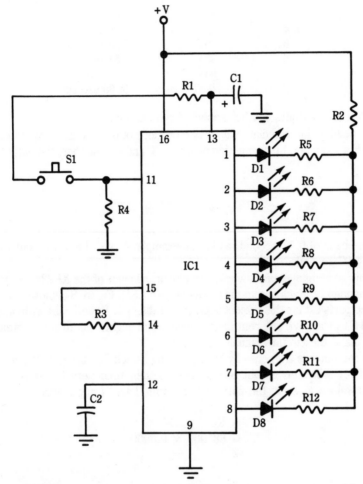

Fig. 8-8. Project 34: sequential countdown flasher.

retriggered via S1. Of course, if you want the circuit to operate with different times, simply change the values of R1 and/or C1.

PROJECT 35: ELECTRONIC CHRISTMAS TREE

Here is a decorative flasher project that is a lot of fun. As described here, the multiple flashing LEDs are arranged to look like a Christmas tree. If you prefer, you can arrange them in any desired pattern.

Basically, this circuit, which is shown in Fig. 8-9, is a XR2240 programmable timer (see project 34) in the astable mode. All eight outputs are used. Note that some of the outputs drive multiple LEDs. You can change the number if you desire, but if any given output has to drive more than three LEDs, a buffer amplifier stage might be required.

**Table 8-4. Parts List for the
Sequential Countdown Flasher Project of Fig. 8-8.**

Component	Description
IC1	XR2240 programmable timer
D1–D8	LED
C1	50-μF, 25-volt electrolytic capacitor
C2	0.01-μF capacitor
R1	680K, 1/4-watt resistor
R2	10K, 1/4-watt resistor
R3	22K, 1/4-watt resistor
R4	1-megohm, 1/4-watt resistor
R5–R12	220-ohm, 1/4-watt resistor
S1	normally open SPST push-switch

Figure 8-10 shows how the LEDs are arranged in the shape of a Christmas tree. You might want to paint a green tree directly on the front panel of the project. Drill holes for the LEDs to go through, like ornaments on the tree.

LED1 is the fastest blinking LED, so it should be placed at the top of the tree. The exact placement of the other LEDs is not critical, but it is a good idea not to put two LEDs from the same output immediately adjacent to one another. Their unison blinking will look a little odd.

You might want to use different colored LEDs for an even more festive appearance.

The parts list for this project is given in Table 8-5. The base flash frequency is determined by the values of resistor R1 and capacitor C1. Feel free to experiment with other values for these components.

The frequency at each output will be determined by this simple formula:

$$F = 1/(2nRC)$$

where:
  R is the resistance of R1,
  C is the capacitance of C1, and
  n is the count value of the appropriate output pin.
  That is;

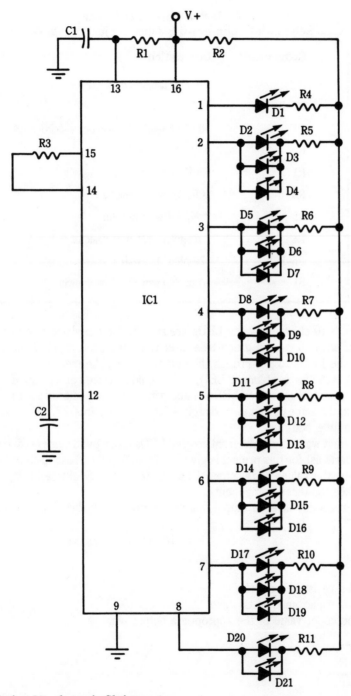

Fig. 8-9. Project 35: electronic Christmas tree.

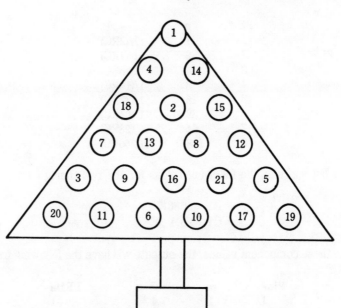

Fig. 8-10. Arrange the LEDs into a suitable Christmas tree shape.

**Table 8-5. Parts List for the
Electronic Christmas Tree Project of Fig. 8-9.**

Component	Description
IC1	XR2240 programmable timer
D1–D21	LED
C1	0.01-μF capacitor
C2	0.01-μF capacitor
R1	680K, 1/4-watt resistor
R2	10K, 1/4-watt resistor
R3	22K, 1/4-watt resistor
R4	180-ohm, 1/4-watt resistor
R5–R10	270-ohm, 1/4-watt resistor
R11	220-ohm, 1/4-watt resistor

Pin#	n	F
1	1	1/(2RC)
2	2	1/(4RC)
3	4	1/(8RC)
4	8	1/(16RC)
5	16	1/(32RC)
6	32	1/(64RC)
7	64	1/(128RC)
8	128	1/(256RC)

The parts list specifies the following values for the two timing components:

$$R1 = 680K$$
$$C1 = 0.1 \ \mu F$$

If you use these component values, the outputs will have the following frequencies:

Pin#	n	F	LEDs
1	1	7.4	1
2	2	3.7	2
			3
			4
3	4	1.8	5
			6
			7
4	8	0.92	8
			9
			10
5	16	0.46	11
			12
			13
6	32	0.23	14
			15
			16
7	64	0.11	17
			18
			19
8	128	0.06	20
			21

The total effect of the finished project is very much like the twinkling lights on a Christmas tree. This project makes a great holiday decoration or gift. You could make a wall-mounted version, or a version to place in the back window of a car. If your tastes are so inclined, you could even wear it on your hat, or as a large brooch. Whatever you do with it, it is a jolly and intriguing decoration sure to arouse people's interest.

PROJECT 36: VARIABLE-RATE FLASHER

The next five projects are all built around flasher ICs. These projects are amazingly simple because a flasher IC includes a complete oscillator IC built into the LED housing. All that is needed externally for basic operation is a power source. For more background information on the flasher LED, refer to Chapter 3.

At the time of this writing, Radio Shack sells a flasher LED. The part number is 276-036. Flasher LEDs are also available from many of the mail-order houses.

When power is applied directly to a flasher LED, it flashes about three times a second. Because all of the circuitry is internal, there is no direct way to control the flash rate. By adding a few external components, however, we can force the flasher LED to operate at a faster rate. The circuit for this variable-rate flasher project is shown in Fig. 8-11. All we've done is add a parallel RC network in the power-supply line. Table 8-6 shows the parts list for this project. There aren't many components in this circuit, but you should experiment with other values.

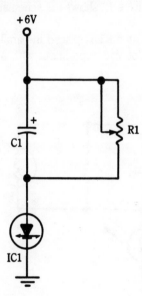

Fig. 8-11. Project 36: variable-rate flasher.

Table 8-6. Parts List for the Variable-Rate Flasher Project of Fig. 8-11.

Component	Description
IC1	flasher IC (Radio Shack#276-036, or similar)
C1	2500-μF, 30-volt electrolytic capacitor
R1	5K potentiometer

The capacitor should be fairly large. Use an electrolytic capacitor of at least 500 μF, or so. The smaller the capacitor is, the faster the flash rate will be. For flash rates higher than 10 to 12 times a second, the human eye will tend to blend the individual flashes together, so the flasher LED will appear to be continuously lit. If capacitor C1 has a value less than 500 μF, the separate flashes will not be visible, which would defeat the whole point of the project.

Potentiometer R1 also has an effect on the flash rate. By using a potentiometer for this resistance, the flash rate can be manually varied.

PROJECT 37: LIGHT-CONTROLLED VARIABLE-RATE FLASHER

This project is very similar to Project 36. The big difference here is that a photoresistor is used to set the flash rate. The light intensity determines how rapidly the LED blinks on and off.

Figure 8-12 shows the schematic diagram for this project. Table 8-7 gives the parts list.

Resistor R1 is placed in parallel with the photoresistor to lower its overall value range. You can experiment with other values for R1 and C1.

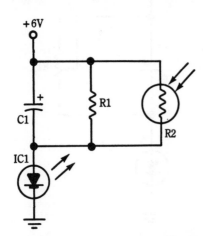

Fig. 8-12. Project 37: light-controlled variable-rate flasher.

Table 8-7. Parts List for the Light-Controlled Variable-Rate Flasher Project of Fig. 8-12.

Component	Description
IC1	flasher IC (Radio Shack#276-036, or similar)
C1	2000-μF, 30-volt electrolytic capacitor
R1	4.7K, 1/4-watt resistor
R2	photoresistor

PROJECT 38: AC-POWERED FLASHER

Most LED flasher circuits are dc powered. In some applications, you might want to drive a flasher LED from an ac source. You could use an ac-to-dc power-supply converter circuit, but that seems like overkill. It seems silly for three-quarters of the project's bulk and cost to be taken up by the power supply.

Figure 8-13 shows a simpler solution. A suitable parts list for this ac-powered flasher project is given in Table 8-8. As you can see, this is a very simple circuit. All that's needed is an ordinary diode and a series-dropping resistor. Note that a ½-watt resistor is called for in the parts list. Do not use a ¼-watt resistor for this application.

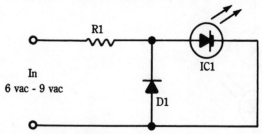

Fig. 8-13. Project 38: ac-powered flasher.

Table 8-8. Parts List for the
AC-Powered Flasher Project of Fig. 8-13.

Component	Description
IC1	flasher IC (Radio Shack #276-036, or similar)
D1	1N4002 diode (or similar)
R1	1K, 1/2-watt resistor

The ac power source should be in the range of 6 volts to 9 volts ac. If you are operating this circuit from the ac power lines, you will need a suitable dropping transformer. A fuse or circuit breaker is also strongly advised.

Of course, whenever you are working with ac power, even at low voltages, follow all necessary precautions. The project must be well-housed. It should be impossible for anyone to come into contact with any conductor carrying ac current.

PROJECT 39: LIGHT-ACTIVATED FLASHER

The circuit shown in Fig. 8-14 is a rather novel flasher project. The LED flashes in response to the ambient lighting level. When the area is dark, the photoresistor (R3) exhibits a very high resistance, holding the flasher LED off. When the ambient light level increases, the flasher LED turns on. The lighting level to some extent controls the flash rate.

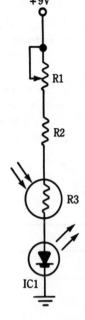

Fig. 8-14. Project 39: light-activated flasher.

Table 8-9. Parts List for the Light-Activated Flasher Project of Fig. 8-14.

Component	Description
IC1	flasher IC (Radio Shack #276-036, or similar)
R1	1K potentiometer
R2	220-ohm, 1/4-watt resistor
R3	photoresistor

Potentiometer R1 serves as a sensitivity control for the photoresistor (R3). A complete parts list for this project appears in Table 8-9.

PROJECT 40: DARK-ACTIVATED FLASHER

This project is along the same lines as Project 39, except it works in exactly the opposite manner. Where Project 39 turned the flasher LED on when the sensor was illuminated, this project turns the flasher LED off when light is striking the photoresistor and on when the sensor is shielded from light, or the environment is dark.

The circuit diagram for this dark-activated flasher project is shown in Fig. 8-15. The parts list appears in Table 8-10.

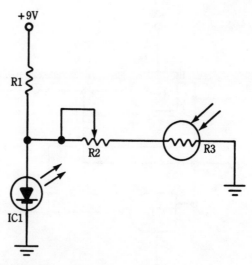

Fig. 8-15. Project 40: dark-activated flasher.

Table 8-10. Parts List for the Dark-Activated Flasher Project of Fig. 8-15.

Component	Description
IC1	flasher IC (Radio Shack #276-036, or similar)
R1	3.9K, 1/4-watt resistor
R2	10K potentiometer
R3	photoresistor

Potentiometer R2 functions as a sensitivity control for the photoresistor (R3), permitting you to manually determine how bright the light must be to turn the flasher LED off.

When the flasher LED is turned on, the flash rate will vary somewhat with the light intensity.

PROJECT 41: TWO-COLOR FLASHER

Any flasher circuit is an eye-catcher, but this project will catch the eye amid even a sea of LED flashers. It flashes two different colors.

Figure 8-16 illustrates the schematic diagram for this project, with the parts list appearing in Table 8-11.

The "secret" of this two-color flasher circuit is the tri-color LED (D1). A tri-color LED contains two back-to-back LEDs in a single housing. One is red and the other is

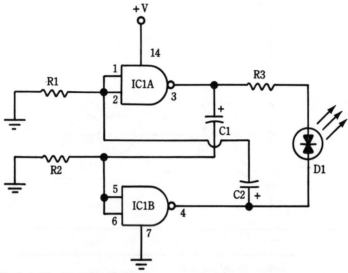

Fig. 8-16. Project 41: two-color flasher.

**Table 8-11. Parts List for the
Two-Color Flasher Project of Fig. 8-16.**

Component	Description
IC1	CD4011 quad NAND gate
D1	Tri-color LED
C1, C2	150-μF, 25-volt electrolytic capacitor
R1, R2	22K, 1/4-watt resistor
R3	330-ohm, 1/4-watt resistor

green. When one of these diodes is forward-biased (lit), the other is reverse-biased (dark), and vice versa.

IC1 is wired as an astable multivibrator circuit. The output of IC1A is always in the opposite logic state as the output of IC1B. When one is low, the other is high. The feedback paths in the circuit force the two outputs to keep reversing states at a regular rate.

The LED (D1) will appear to be continuously lit, but it will keep changing color, from red to green and back again. These oscillating colors really draw the eye.

Experiment with other component values for resistors R1 and R2 and capacitors C1 and C2. These components control the timing of the astable multivibrator. If equal components are used in both halves of the circuit (C1 = C2 and R1 = R2), the red and green light times will be more or less equal.

The component values in the parts list were selected to give a fairly low frequency. If a higher frequency is set up (above about 10 to 12 Hz), the two colors will blend into each other, and the LED will appear to be continuously lit with a yellow glow.

For variety, you could replace resistors R1 and R2 with a single potentiometer. Ground the wiper. This modification allows you to "detune" the direction to favor either the red light or the green light. If the potentiometer is set at its midpoint, each color turns on for the same amount of time. Moving the wiper in one direction causes the red light to last longer than the green light. Adjusting the potentiometer in the other direction produces just the opposite effect, of course.

For more information on tri-color LEDs and how they work, refer to Chapter 3.

PROJECT 42: HIGH-INTENSITY LED FLASHER

The more current there is flowing through an LED, the brighter it glows. If too much current flows through an LED, however, the semiconductor junction can be destroyed. This is the purpose of the current-dropping series resistor found in virtually all LED circuits.

So what can you do if you need to use an LED, and it just isn't bright enough?

One possible solution is illustrated in Fig. 8-17. The parts list for this project appears in Table 8-12.

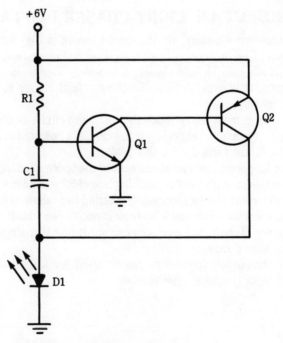

Fig. 8-17. Project 42: high-intensity LED flasher.

Fig. 8-12. Parts List for the High-
Intensity LED Flasher Project of Fig. 8-17.

Component	Description
Q1	npn transistor (2N3904, or similar)
Q2	pnp transistor (2N3906, or similar)
D1	LED
C1	0.05-μF capacitor
R1	330K, 1/4-watt resistor

In this circuit, brief, high-current pulses are applied to the LED (D1). The pulses are too short to burn out the semiconductor junction within the LED, but they are spaced closely enough so that the eye cannot distinguish between the individual flashes. The LED will appear to be continuously lit at a higher than normal intensity.

If you use battery power, you should be aware that this circuit is a current hog. Batteries won't last nearly as long in this circuit as they will in other "normal" flasher circuits.

PROJECT 43: LIGHT-CHASER DISPLAY

For a really impressive display, try the circuit shown in Fig. 8-18. By physically arranging the fifteen LEDs in numerical order, an illusion of movement can be achieved. The effect is commonly called a light chaser. It is widely used in advertising displays, and for good reason. It is eye-catching. Watching a light chaser is fascinating, in a semihypnotic sort of way.

The simplest arrangement is to position the LEDs in a circle, as shown in Fig. 8-19. The circle will appear to revolve, especially if the circuit is operated in a darkened area.

Table 8-13 shows the parts list for this project.

Potentiometer R2 permits manual adjustment of the speed. You might want to eliminate this control and use a single fixed resistor in place of R2 and R3 in some applications.

The sequence rate can also be changed by altering the values of capacitors C2 and C3. These two capacitors are in parallel, so their capacitances add. If you have a single nonpolarized capacitor of the right value, you can substitute that single unit in place of the two capacitors shown here.

Almost any common npn transistors can be used for Q1 through Q5. All five transistors should be of the same type number.

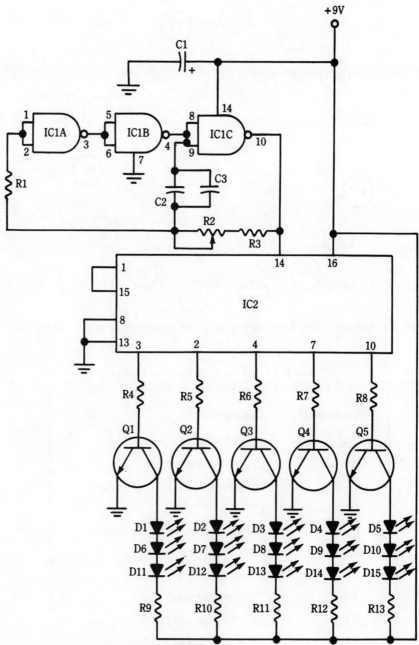

Fig. 8-18. Project 43: light-chaser display.

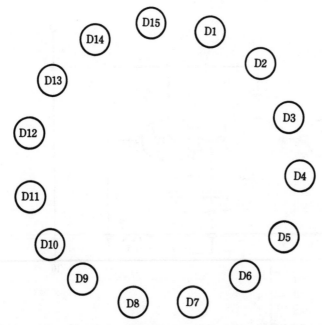

Fig. 8-19. The best effects in a light-chaser display can be achieved by arranging the LEDs in a circle.

Table 8-13. Parts List for the Light-Chaser Display Project of Fig. 8-18.

Component	Description
IC1	CD4011 quad NAND gate
IC2	CD4017 decade counter
Q1–Q5	npn transistor (2N3904, or similar)
D1–D15	LED
C1	250-μF, 25-volt electrolytic capacitor
C2, C3	0.47-μF capacitor
R1	1-megohm, 1/4-watt resistor
R2	10K potentiometer
R3	12K, 1/4-watt resistor
R4–R8	270-ohm, 1/4-watt resistor
R9–R13	10K, 1/4-watt resistor

Chapter 9

Test Equipment and Measurement Devices

Electrical quantities such as voltage, resistance, and current cannot be detected and measured directly by human senses. Of course, if you touch a wire carrying a high ac current, you're going to know it, but that is scarcely a very practical way to determine what is happening within the circuit.

This chapter features a number of projects for building your own test equipment and measuring devices, using optoelectric components.

PROJECT 44: BARGRAPH DISPLAY

Digital readouts are very useful in many measurement applications. They are easy to read, and are precise and unambiguous. In some applications, however we don't really need the precision. We just want an easy way to read general indication of which range the signal is in.

If you don't want to use an old-fashioned panel meter, you can use a bargraph. A bargraph is made up of a series of LEDs (or other indicating devices) arranged in a row, or bar. The more LEDs that are lit, the higher the input signal level is. The idea of the bargraph is illustrated in Fig. 9-1.

The simplest way to create a bargraph voltmeter is to use multiple-voltage comparators. This is the approach used in this project. The circuit for a four-stage voltage bargraph is shown in Fig. 9-2. A typical parts list for this project is given in Table 9-1.

The number of lit LEDs gives an indication of the approximate input voltage. As the input voltages increases, more of the comparator outputs will go high, lighting up more of the LEDs. Unlike an ordinary analog panel meter, the bargraph is easy to read, even at a quick glance. If LEDs are used as the indicating devices, as in this project, the value can even be read in the dark.

117

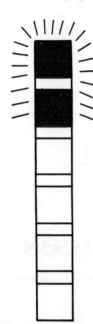

Fig. 9-1. The simplest multi-LED readout is the bargraph.

The circuit as shown here uses all four sections of a single LM339 quad comparator IC (see Fig. 9-3). There is nothing critical about the number of stages. Four stages are used here. You can increase the bargraph up to eight stages by using a second LM339 quad comparator IC.

The value of resistor R1 determines the sensitivity of the circuit. You could use a potentiometer here as a calibration control, if that is appropriate to your specific application.

Resistors R2 through R5 set up the measurement ranges. Experiment with different values. By making these resistors equal, as indicated in the parts list, the signal is broken up into equal steps per LED. If your application calls for a weighted or nonlinear scale of some kind, you can substitute unequal values for these resistors.

Resistors R6 through R9 simply protect the LEDs from excessive current flow. Their values determine the brightness of their associated LEDs.

This is a simple, but versatile project with hundreds of possible applications. The only precaution to bear in mind when using this circuit is that the input signal is not referenced directly to ground. The input voltage is connected across the two points in the diagram marked V_{in}. Do not ground one end of the input signal.

PROJECT 45: TEN-STEP BARGRAPH DISPLAY

The simple bargraph display circuit of Project 44 is functional, but really only suitable for fairly low-resolution applications. Because a separate comparator stage with dropping resistors is required for each additional output LED, the system can quickly become unwieldy.

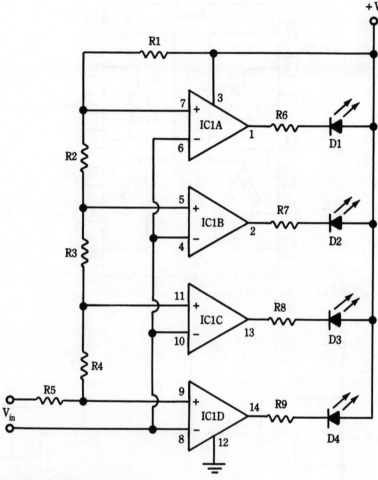

Fig. 9-2. Project 44: bargraph display.

Table 9-1. Parts List for the Bargraph Display Project of Fig. 9-2.

Component	Description
IC1	LM339 quad comparator
D1–D4	LED
R1	68K, 1/4-watt resistor
R2–R5	2.2K, 1/4-watt resistor
R6–R9	470-ohm, 1/4-watt resistor

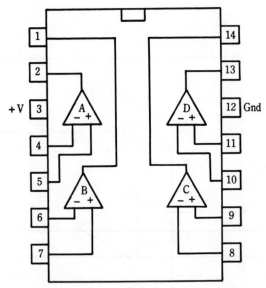

Fig. 9-3. Project 44 is built around an LM339 quad comparator IC.

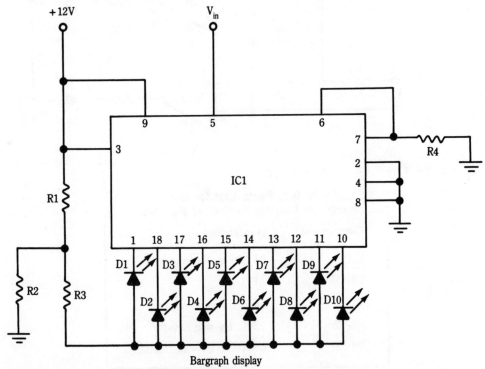

Fig. 9-4. Project 45: ten-step bargraph display.

**Table 9-2. Parts List for the Ten-
Step Bargraph Display Project of Fig. 9-4.**

Component	Description
IC1	LM3914 dot/bar driver
D1–D10	LED
R1	12K, 1/4-watt resistor
R2	10K, 1/4-watt resistor
R3	180-ohm, 1/4-watt resistor
R4	1.5K, 1/4-watt resistor

Several specialized ICs have been developed specifically for use in bar and dot-graph displays. One such device is the LM3914. This chip can drive up to ten LEDs in either a bargraph or dot-graph mode.

Figure 9-4 shows the basic bargraph circuit. A parts list for this project appears in Table 9-2. Using the component values listed here, the full-scale range for this bargraph is 1.2 volts. Each additional LED that lights up indicates a step of 0.12 volt.

PROJECT 46: TEN-STEP DOT-GRAPH DISPLAY

In a bargraph, all of the LEDs up to the highest indicated value are lit. In a dot-graph, only a single LED is lit at any time. All lower valued LEDs are turned off. Bargraphs and dot-graphs are really interchangeable in most applications. The choice depends primarily on personal preference.

A circuit for a ten-step dot-graph display project is shown in Fig. 9-5. Table 9-3 shows the parts list.

Note how similar this project is to Project 44. The only difference is the connection to pin #9. This pin selects the LM3914's operating mode. If pin #9 is connected to V+ (pin #5), the IC functions in the bargraph mode. To operate it in the dot-graph mode, ground pin #9. If pin #9 is left floating, internal circuitry will pull it down to ground level, resulting in the dot-graph mode of operation. I recommend making the actual connection to a solid external ground, just to be sure. Personally, I'm usually hesitant about leaving control pins floating, even when the manufacturer says it is okay. A solid external connection is even more reliable.

PROJECT 47: VARIABLE-RANGE BARGRAPH DISPLAY

A simple modification to the basic LM3914 bargraph display circuit permits you to adjust the range manually. The modified circuit used in this project is illustrated in Fig. 9-6.

A suitable parts list for this project appears in Table 9-4.

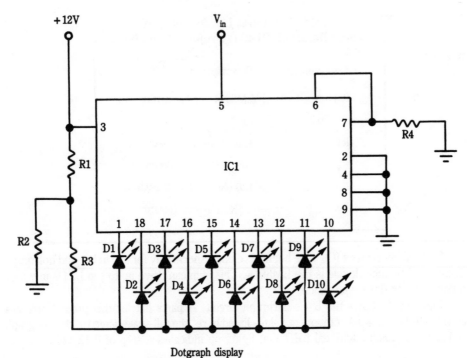

Dotgraph display

Fig. 9-5. Project 46: ten step dot-graph display.

Table 9-3. Parts List for the Ten-Step Dot-Graph Display Project of Fig. 9-5.

Component	Description
IC1	LM3914 dot/bar driver
D1–D10	LED
R1	12K, 1/4-watt resistor
R2	10K, 1/4-watt resistor
R3	180-ohm, 1/4-watt resistor
R4	1.5K, 1/4-watt resistor

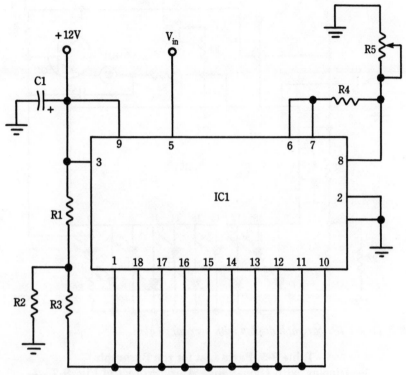

Fig. 9-6. Project 47: variable-range bargraph display.

Table 9-4. Parts List for the Variable-Range Bargraph Display Project of Fig. 9-6.

Component	Description
IC1	LM3914 dot/bargraph display driver
D1–D10	LED
C1	10-μF, 25-volt electrolytic capacitor
R1	12K, 1/4-watt resistor
R2	10K, 1/4-watt resistor
R3	180-ohm, 1/4-watt resistor
R4	1.5K, 1/4-watt resistor
R5	10K potentiometer

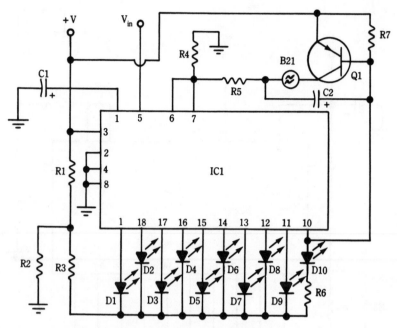

Fig. 9-7. Project 48: bargraph display with over-range alert.

**Table 9-5. Parts List for the Bargraph
Display with Over-Range Alert Project of Fig. 9-7.**

Component	Description
IC1	LM3914 dot/bargraph display driver
Q1	pnp transistor (2N3906, or similar)
D1–D10	LED
C1	10-μF, 25-volt electrolytic capacitor
C2	100-μF, 25-volt electrolytic capacitor
R1	12K, 1/4-watt resistor
R2	10K, 1/4-watt resistor
R3	180-ohm, 1/4-watt resistor
R4	1.5K, 1/4-watt resistor
R5	680-ohm, 1/4-watt resistor
R6	100-ohm, 1/4-watt resistor
R7	1K, 1/4-watt resistor
BZ1	piezoelectric buzzer

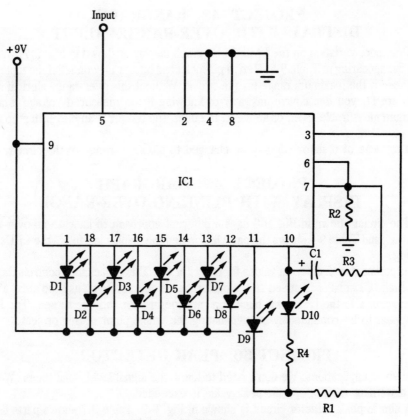

Fig. 9-8. Project 49: bargraph display with blinking over-range.

Table 9-6. Parts List for the Bargraph
── Display with Blinking Over-Range Project of Fig. 9-8. ──

Component	Description
IC1	LM3914 dot/bargraph display driver
D1–D10	LED
C1	100-μF, 35-volt electrolytic capacitor
R1	1K, 1/4-watt resistor
R2	1.2K, 1/4-watt resistor
R3	620-ohm, 1/4-watt resistor
R4	100-ohm, 1/4-watt resistor

PROJECT 48: BARGRAPH
DISPLAY WITH OVER-RANGE ALERT

One more variation on the LM3914 bargraph display is shown in Fig. 9-7, with the parts list appearing in Table 9-5. Here a piezoelectric buzzer is sounded if the input voltage exceeds the maximum range for the circuit. Without this over-range alert, if all ten LEDs are lit, you don't have any way of knowing if the measured voltage is at the maximum measurable level, or above it. In effect, the buzzer is an eleventh step in the readout.

The value of resistor R5 may be changed to alter the range of the circuit.

PROJECT 49: BARGRAPH
DISPLAY WITH BLINKING OVER-RANGE

The circuit shown in Fig. 9-8 uses a different approach to indicate an over-range condition, and Table 9-6 shows the parts list. When the display is "full," the LEDs start blinking.

Experiment with different values for capacitor C1. This capacitance controls the flash rate. Make C1 smaller to speed up the blink rate. Of course, if you use too small a value for capacitor C1, the LEDs will blink at a rate too fast for the eye to see. The LEDs will appear to be continuously lit, eliminating the whole point of this project.

PROJECT 50: PEAK DETECTOR

In some applications, we don't need to know the signal level at all times. We just need a warning when a specific peak value is exceeded.

A simple peak detector circuit is shown in Fig. 9-9. Table 9-7 shows parts list for this project.

When the amplitude of the input signal exceeds the peak value set via potentiometer R1, the LED (D2) lights up. When the input signal drops back down below the peak level, the LED goes dark again.

This project can be used in many applications. For example, you could use it to monitor a loudspeaker, a PA system, or a tape recorder input to determine when the signal gets too high, which could result in distortion. In all of these sample applications, a few brief flashes can be ignored, but if the LED lights up frequently, or stays lit more than a fraction of a second, the signal amplitude should be reduced.

PROJECT 51: FREQUENCY DETECTOR

This project can be used to determine if a signal is at a specific desired frequency. The circuit, shown in Fig. 9-10, is built around two 567 tone-decoder ICs. Table 9-8 shows the complete parts list for this project.

The frequency of interest is set via potentiometers R4 and R6. A precise calibration signal known to be at the desired frequency is quite helpful in working with this circuit.

When the input signal is at, or very close to, the desired frequency, LED D3 will glow. Otherwise LED D1 and/or LED D2 will light up.

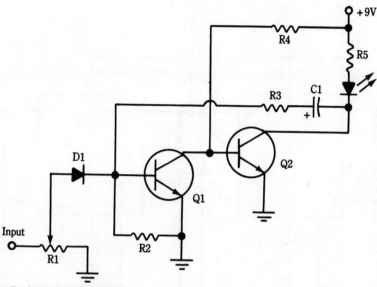

Fig. 9-9. Project 50: peak detector.

**Table 9-7. Parts List for the
Peak Detector Project of Fig. 9-9.**

Component	Description
Q1, Q2	npn transistor (HEP-55, ECG-123A, or similar)
D1	diode (1N4148, or similar)
D2	LED
C1	100-μF, 25-volt electrolytic capacitor
R1	1K potentiometer
R2	47K, 1/4-watt resistor
R3	22K, 1/4-watt resistor
R4	2.2K, 1/4-watt resistor
R5	390-ohm, 1/4-watt resistor

This project is useful for applications that require a signal source be tuned. Adjust the signal source frequency until LED D3 lights up and LED D1 and LED D2 go dark.

PROJECT 52: PHASE AND FREQUENCY COMPARATOR

Frequency meters are relatively complex and expensive devices. Often they aren't really necessary. This is especially true in applications where a variable signal must be

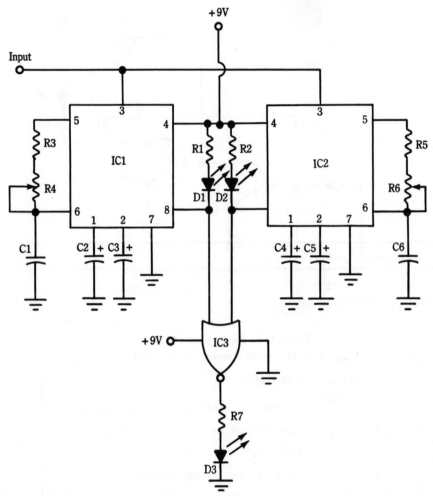

Fig. 9-10. Project 51: frequency detector.

tuned to a specific standard. In such applications, it is often more convenient to use a frequency-comparator circuit.

The circuit shown in Fig. 9-11 is a simple but effective means of comparing frequencies. The parts list for this project is given in Table 9-9.

This circuit is actually a phase detector. It checks the input signal to determine if it is in phase with the reference signal, that is, it checks to see if each cycle begins and ends at the same time.

Of course, if the frequencies of the two signals are not identical, they can't possibly stay in phase with each other for more than a fraction of a second. The effect is the same as comparing the frequencies.

Table 9-8. Parts List for the
Frequency Detector Project of Fig. 9-10.

Component	Description
IC1, IC2	567 tone decoder
IC3	CD4001 quad NOR gate
D1, D2, D3	LED
C1, C6	0.1-μF capacitor
C2, C4	5-μF, 25-volt electrolytic capacitor
C3, C5	1-μF, 25-volt electrolytic capacitor
R1, R2, R7	470-ohm, 1/4-watt resistor
R3, R5	2.7K, 1/4-watt resistor
R4, R6	15K, potentiometer

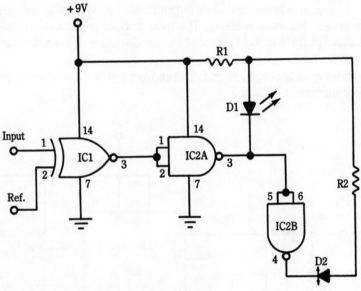

Fig. 9-11. Project 52: phase and frequency detector.

For a positive output indication from this circuit the input signal must be at the same frequency as the reference signal, and the two signals must be in phase with one another.

Because digital gates make up the bulk of this project, both input signals should be either square or rectangle waves. Other (analog) waveforms might cause unreliable circuit operation.

Table 9-9. Parts List for the Phase and Frequency Detector Project of Fig. 9-11.

Component	Description
IC1	CD4070 quad X-OR gate (exclusive OR)
IC2	CD4011 quad NAND gate
D1, D2	LED
R1, R2	470-ohm, 1/4-watt resistor

If the input signal exactly matches the reference signal (both frequency and phase), then LED D2 will light up, while LED D1 will remain dark. If the input signal is at a different frequency, or is out of phase with the reference signal, LED D2 will be dark, and LED D1 will glow. If you don't want to bother with the double indicators, you can eliminate LED D1 and resistor R1 from the circuit.

PROJECT 53: SOUND-LEVEL INDICATOR

If you ever need to know if the signal output from an audio amplifier exceeds a specific level, this project will come in handy. The most obvious application for this project is to detect signal peaks that could potentially cause distortion or even damage a speaker cone.

The circuit is fairly simple, as you can see from Fig. 9-12. A suitable parts list for this project appears in Table 9-10.

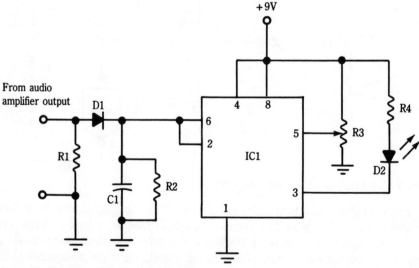

Fig. 9-12. Project 53: sound-level indicator.

**Table 9-10. Parts List for the
Sound-Level Indicator Project of Fig. 9-12.**

Component	Description
IC1	555 timer
D1	diode (1N4148, or similar)
D2	LED
C1	4.7-μF, 25-volt electrolytic capacitor
R1	8.2-ohm resistor (wattage to suit amplifier output)
R2	33K resistor
R3	10K potentiometer
R4	330-ohm, 1/4-watt resistor

The 555 timer (IC1) functions as a Schmitt trigger. When the signal exceeds a specific level, the LED (D2) lights up.

Potentiometer R3 sets the sensitivity of the circuit. This control is used to adjust the signal level that will turn the LED on.

The value of resistor R1 should match the output impedance of the amplifier. In some cases, this circuit can be placed in parallel with the speaker or other output device. If you do this, be sure to account for what happens when resistances (and impedances) are connected in parallel. Remember, parallel resistances (impedances) are always less than any of the individual component values in the parallel combination.

This circuit is designed for use with fairly low-power audio amplifiers. If you want to monitor the output level of a high-wattage amplifier, you need to include some kind of power dropping network so as not to damage the components (especially IC1) in this project.

PROJECT 54: AUDIO WATTMETER

The circuit shown in Fig. 9-13 gives a bargraph indication of the power output of an audio amplifier. Using the component values given in Table 9-11, this audio wattmeter will read power levels up to 100 watts, in the following steps:

Highest Lit LED	Wattage
D1	0.2 watt
D2	0.4 watt
D3	0.8 watt
D4	1.6 watts
D5	3.2 watts
D6	6.4 watts

D7	12.8 watts
D8	25.6 watts
D9	51.2 watts
D10	102.4 watts

Note that the scale is exponential rather than linear. This is because of the way the human ear hears changes in volume. A doubling of the wattage does not result in a doubling of the volume. For the signal volume to be doubled, the wattage must be increased by a factor of 10. Thus, 100 watts is only twice as loud as 10 watts.

In the parts list (Table 9-11), it is assumed that the amplifier's output impedance is rated for a load of 8 ohms. If your audio amplifier has a different output impedance, you might need to change the value of resistor R1. For a 4-ohm load impedance, use a 10K resistor for R1. Some audio amplifiers have an output impedance of 16 ohms. In this case, R1 should have a value of 33K.

PROJECT 55: WIDE-RANGE VU METER

VU meters are used in many audio applications, especially where recording is involved. *VU* stands for *volume units*. A volume unit is a somewhat arbitrary unit that can be used for comparing audio-signal levels.

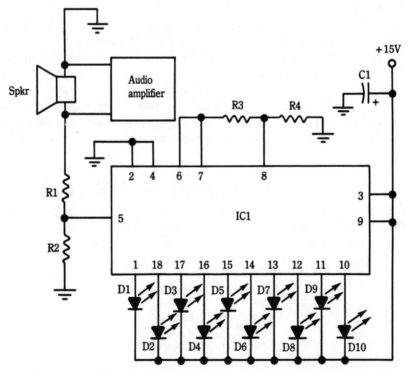

Fig. 9-13. Project 54: audio wattmeter.

**Table 9-11. Parts List for the
Audio Wattmeter Project of Fig. 9-13.**

Component	Description
IC1	LM3915 dot/bargraph display driver
D1–D10	LED
C1	10-μF, 35-volt electrolytic capacitor
R1	18K resistor*
R2	10K resistor*
R3	390-ohm, 1/4-watt resistor
R4	2.7K, 1/4-watt resistor

*wattage to suit amplifier output

A circuit for an LED-readout VU meter with a wider than usual range is shown in Fig. 9-14. Table 9-12 shows the parts list for this project.

The input signal should have a maximum amplitude of 1.25 volts. An external limiting stage might be required in some applications.

The volume units are displayed by the LEDs in bargraph fashion. The highest lit LED indicates the current VU value:

Highest Lit LED	VU
D1	−40 dB
D2	−37 dB
D3	−34 dB
D4	−31 dB
D5	−28 dB
D6	−25 dB
D7	−22 dB
D8	−19 dB
D9	−16 dB
D10	−13 dB
D11	−10 dB
D12	−7 dB
D13	−5 dB
D14	−3 dB
D15	−1 dB
D16	0 dB
D17	+1 dB
D18	+2 dB
D19	+3 dB

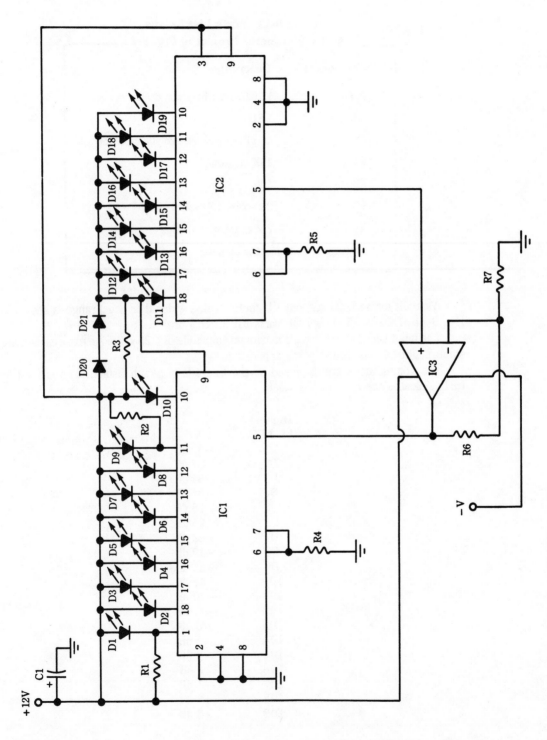

Fig. 9-14. Project 55: wide-range VU meter.

**Table 9-12. Parts List for the Wide-
Range VU Meter Project of Fig. 9-14.**

Component	Description
IC1	LM3915 dot/bargraph display driver
IC2	LM3916 dot/bargraph display driver
IC3	op amp (307, or similar)
D1–D19	LED
D20, D21	diode (1N4148, or similar)
C1	10-μF, 35-volt electrolytic capacitor
R1	12K, 1/4-watt resistor
R2	22K, 1/4-watt resistor
R3, R4, R5	1K, 1/4-watt resistor
R6	330K, 1/4-watt resistor
R7	62K, 1/4-watt resistor

In most applications, the 0-dB point is considered the maximum signal level without an unacceptable amount of distortion. Brief peaks above 0 dB are okay, but they should be very short and fairly rare.

For each visibility, you might want to make D16 through D19 a different color than D1 through D15. Generally, green will be used for values below 0 dB, while red is used for 0 dB and up. You might also want to position LEDs D16 through D19 in some special way to make them more readily visible at a glance.

PROJECT 56: ALTERNATE PEAK DETECTOR

We will conclude this chapter with yet another simple peak-detector project. This one is extremely simple, consisting of four readily available components.

As you can see in the circuit diagram of Fig. 9-15, this project is built around an op amp. Any op-amp IC can be used, so long as it suits the needs of your specific application. For serious applications requiring high precision, you should obviously use a high-grade, low-noise op amp. For less critical applications, you can use a less expensive chip, such as the popular 741 op amp.

Note that a single-polarity power supply is being used here. In this application, we can get away with this, even though most op-amp ICs normally require a dual-polarity supply. The only restriction is that the input voltage should be positive only. Do not use negative polarity or dual polarity signals with this circuit.

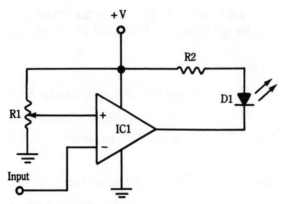

Fig. 9-15. Project 56: peak detector.

**Table 9-13. Parts List for the
Peak Detector Project of Fig. 9-15.**

Component	Description
IC1	op amp
D1	LED
R1	100K potentiometer
R2	470-ohm, 1/4-watt resistor

The parts list for this project, simple as it is, is given in Table 9-13.

Potentiometer R1 sets the reference voltage for the comparator. If the input voltage exceeds this reference value, the LED (D1) lights up. As long as the input voltage is less than the reference voltage set by R1, the LED remains dark.

Chapter 10

Game Projects

The projects in this chapter are all just for fun. With them you can build your own electronic games.

Some of these projects duplicate or aid in the play of standard games. Others are possible only with electronics.

PROJECT 57: ELECTRONIC DICE

In many games, each player's choice of moves is determined by a roll of the dice. Why not build a set of electronic dice?

In this project, a random-number generator selects a value from 1 to 6. This number is displayed on a set of LEDs arranged in the traditional die face pattern.

Figure 10-1 shows the circuitry for a single electronic die. The parts list for this project appears in Table 10-1.

Of course, if the game you want to play calls for two dice, you can just build two of these projects. It might make sense to build two electronic die circuits in a common housing. The displays (LEDs) can be mounted side by side. One dual-pole pushbutton switch can be used to control both circuits (in place of two separate SPST switches—S1). Naturally, a common power supply would also be used for the two circuits.

In this circuit, a CD4017 decade counter chip (IC2) is connected as a modulo-six cycling counter. The gates of IC3 and IC4 determine which LEDs will be lit for each count.

The timer (IC1) and its associated components form a clock, supplying pulses to advance the counter. When pushbutton switch S1 is depressed (closed), the clock pulses can get to the counter. When the switch is released (open), the clock signal is broken off, and the counter holds its last value.

While the clock is feeding the counter (ie, S1 is closed), all seven LEDs appear to be continuously lit. This is because the clock rate is quite high. There is no way for anyone to predict what the displayed value will be when the switch is released.

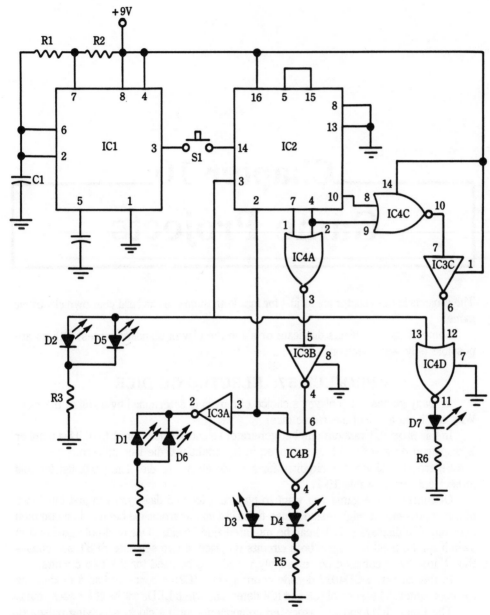

Fig. 10-1. Project 57: electronic dice.

Figure 10-2 illustrates how the seven output LEDs (D1–D7) should be mounted to simulate traditional dice face patterns. Follow the numbering carefully, or you will get some very odd-looking patterns. The count will be correct, but the display won't resemble an ordinary die.

Table 10-1. Parts List for the
Electronic Dice Project of Fig. 10-1.

Component	Description
IC1	555 timer
IC2	CD4017 decade counter
IC3	CD4049 hex inverter
IC4	CD5001 quad NAND gate
D1–D7	LED
C1, C2	0.1-μF capacitor
R1, R2	1K, 1/4-watt resistor
R3–R6	330-ohm, 1/4-watt resistor
S1	normally open SPST push-switch

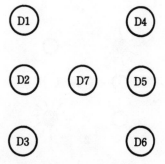

Fig. 10-2. The seven output LEDs should be mounted to simulate the traditional die face pattern.

Table 10-2 summarizes the LEDs that are lit for each individual count value. The six possible dice face patterns are illustrated in Fig. 10-3 for your convenience.
Note that certain LEDs are always lit up in pairs:

D1 and D6
D2 and D5
D3 and D4

This simplifies the gating and wiring for the circuit. These paired diodes are simply connected in parallel, with a common current-dropping series resistor.

**Table 10-2. Summary of LED
Action in the Electronic Dice Project.**

Component	Description						
Counter	LEDs lit						
Output	1	2	3	4	5	6	7
1	.	.	.	.	.	.	X
2	X	.	.	.	.	X	.
3	X	.	.	.	.	X	X
4	X	.	X	X	.	X	.
5	X	.	X	X	.	X	X
6	X	X	X	X	X	X	.

X = LED on
. = LED off

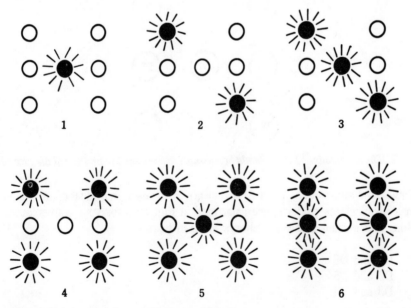

Fig. 10-3. By lighting the correct LEDs, any standard die face pattern can be displayed.

PROJECT 58: "WHO'S FIRST" GAME

This is an exciting game that allows three people to pit their reflexes against one another.

The circuit diagram is broken into two parts and is shown in Fig. 10-4 and Fig. 10-5. The parts list for this project appears in Table 10-3.

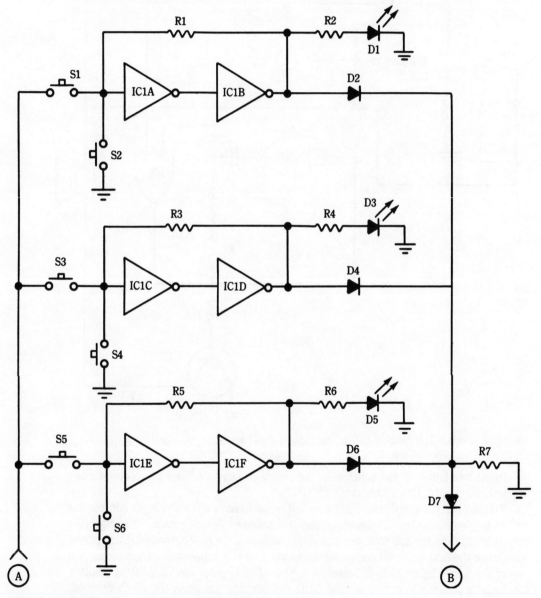

Fig. 10-4. Project 58A: first half of the "who's first" game.

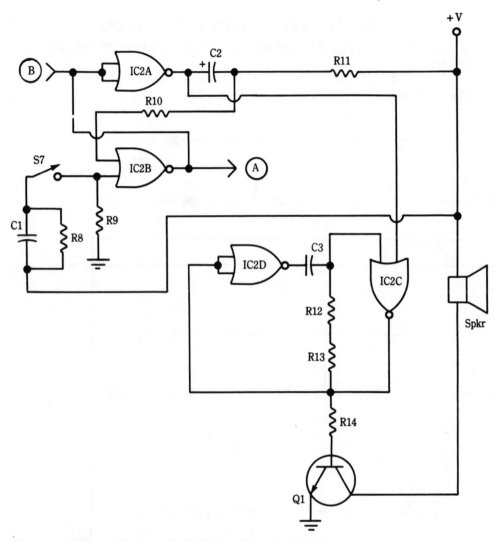

Fig. 10-5. Project 58B: second half of the "who's first" game.

In the two halves of the schematic, the points marked "A" are connected to one another, as are the two points marked "B."

To start the game, switch S7 is closed. It helps to have a fourth player control this switch to give a true test of the other players' reflexes. When switch S7 is closed, a tone is heard from the speaker, and the player switches are now activated. Each player tries to be the first to hit his push-switch. As soon as this happens the tone stops, and one of the LEDs lights up to indicate the winner.The appropriate CLEAR pushbutton can then be pressed to extinguish the LED and prepare the game for another round.

**Table 10-3. Parts List for the
"Who's First" Game Project of Figs. 10-4 and 10-5.**

Component	Description
IC1	CD4049 hex inverter
IC2	CD4001 quad NOR gate
Q1	npn transistor (2N3904, 2N2222, or similar)
D1, D3, D5	LED
D2, D4, D6, D7	diode (1N4148, or similar)
C1, C3	0.01-μF capacitor
C2	1-μF, 25-volt electrolytic capacitor
R1, R3, R5	10 megohm, 1/4-watt resistor
R2, R4, R6	1K, 1/4-watt resistor
R7, R9, R10	100K, 1/4-watt resistor
R8	1 megohm, 1/4-watt resistor
R11	470K, 1/4-watt resistor
R12	22K, 1/4-watt resistor
R13	8.2K, 1/4-watt resistor
R14	10K, 1/4-watt resistor
S1–S6	normally open SPST momentary pushbutton switch
S7	SPST switch
Spkr	small speaker

Each player has his own stop switch, clear switch, and LED:

Player	Stop	Clear	LED
1	S1	S2	D1
2	S3	S4	D3
3	S5	S6	D5

Control circuits for other players can easily be added in parallel with the three shown in Fig. 10-4.

PROJECT 59: SCOREKEEPER

A natural application for digital electronics is a counter. It is easy to adapt a counter to serve as a scorekeeping device for games and sporting events. When a player scores, the appropriate button is pushed to increment the counter.

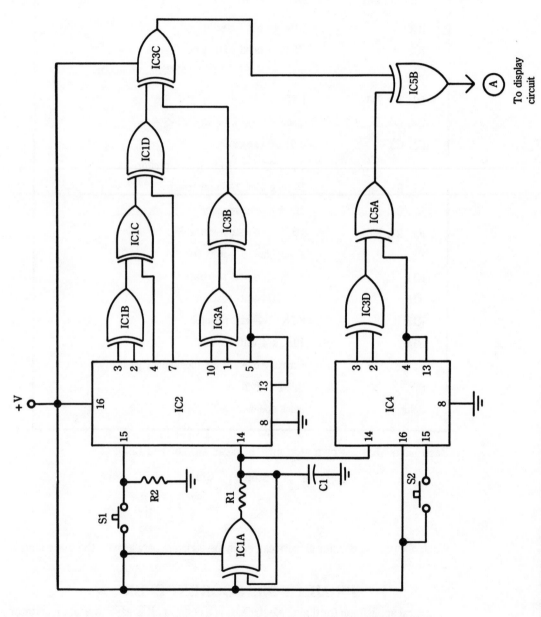

Fig. 10-6. Project 59A: first half of the scorekeeper.

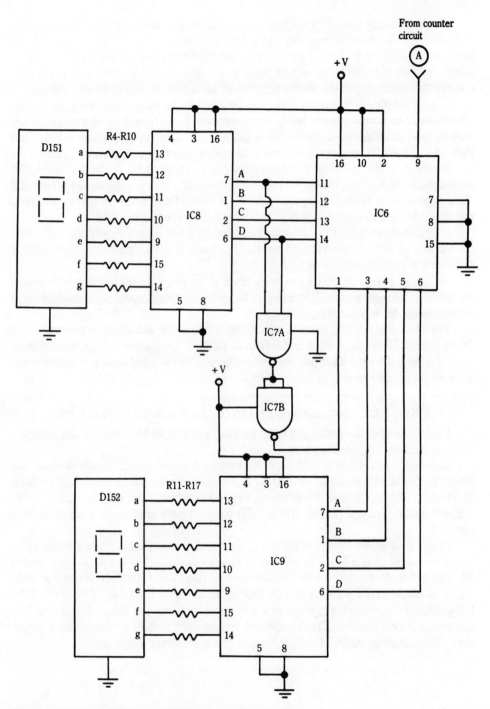

Fig. 10-7. Project 59B: second half of the scorekeeper.

However, many games do not score with simple increments of one. For example, in football a touchdown scores six, and a field goal scores three. The operator could manually press the switch six or three times, but this would be a nuisance at best. There is also the risk of losing count of how many times the switch has been pressed so far. Clearly this could defeat the entire purpose of an automatic scorekeeping circuit.

A good solution is to use multiple counters within the single scorekeeping circuit. The CD4017 can come in very handy here, because it can be set up as a single cycle-manual reset counter. When the button is pushed, the single cycle counter is reset and then run to feed the appropriate number of pulses into the main (score) counter.

Figure 10-6 and 10-7 show a complete football scoring unit for one team. Table 10-4 shows the parts list. The circuit is broken into two parts for convenience in presenting the diagrams here. The points marked "A" in each half of the circuit should be connected together. The same power supply is used for both halves of the circuit.

In operation, switch S1 is momentarily closed for a touchdown (6 points), and switch S2 is briefly closed for a field goal (3 points). IC1 and IC2 are simply debouncing circuits to keep the input signals clean.

As shown here, the circuit can keep track of scores from 00 to 99, which should be sufficient for most games. If you need a larger count range, simply add another stage as you would for any counter.

For a practical system, you will need to build two of the scorekeeper projects—one for each team. Both scorekeeper circuits can be built into a single housing for convenience.

Of course, the same basic principle utilized here can be adapted for appropriate scoring increments for almost any game or sport.

PROJECT 60: ELECTRONIC ROULETTE WHEEL

Roulette is a popular game, and it can be a lot of fun. It can be simulated electronically using a random-number generator.

Ordinarily, roulette is played on a mechanical roulette wheel. In this project, the wheel is simulated with a ring of LEDs. The players try to guess which LED will be lit when the counter stops. If you are so inclined you could also use this circuit as an "ESP" tester. Can you predict Which LED will light with greater accuracy than pure chance?

Figure 10-8 shows the circuit for this project, and the parts list is given in Table 10-5.

Three sections of a 7404 hex inverter are used as a clock. When a pushbutton switch S1 is closed momentarily, the circuit starts oscillating. The frequency starts out fairly high (the actual base frequency can be altered by changing the value of capacitor C1), but gradually slows down and stops at a rate determined by the value of capacitor C2. Increasing the capacitance of the component causes the "wheel" to "spin" for a longer time. A smaller capacitor, of course, will produce a shorter sequence.

**Table 10-4. Parts List for the
Scorekeeper Project of Figs. 10-6 and 10-7.**

Component	Description
IC1, IC3, IC5	CD4070 quad X-OR gate
IC2, IC4	CD4017 decade counter
IC6	CD4518 dual BCD counter
IC7	CD4011 quad NAND gate
IC8, IC9	CD4511 BCD to 7-segment decoder
DIS1, DIS2	common-cathode 7-segment LED display
C1	0.005-μF capacitor
R1	4.7K, 1/4-watt resistor
R2, R3	1K, 1/4-watt resistor
R4–R17	330-ohm, 1/4-watt resistor
S1, S2	normally open SPST pushbutton switch

As a bonus, a clicking sound effect is added each time the count is incremented. This improves the simulation of a real mechanical roulette wheel. If you don't want the sound effect, you can simply eliminate the following components from your project:

1. IC1D
2. Q1
3. R4
4. C3
5. Spkr

If you do include the sound-effect feature, almost any "garden variety" npn transistor may be used for Q1. The specifications are not at all critical.

The counter drives a 4-to-16 line decoder (IC3). Sixteen LEDs are used to display the output of this chip. Only one LED will ever be lit at any given time. To simulate a roulette wheel, the LEDs should be mounted in numerical order in a circular pattern. As each LED is lit and extinguished in turn, the display will give the effect of a spinning wheel.

PROJECT 61 "QUICK-DRAW" GAME

This "Quick-Draw" game is a great way to test your reflexes. The circuit is shown in Fig. 10-9 and the parts list appears in Table 10-6.

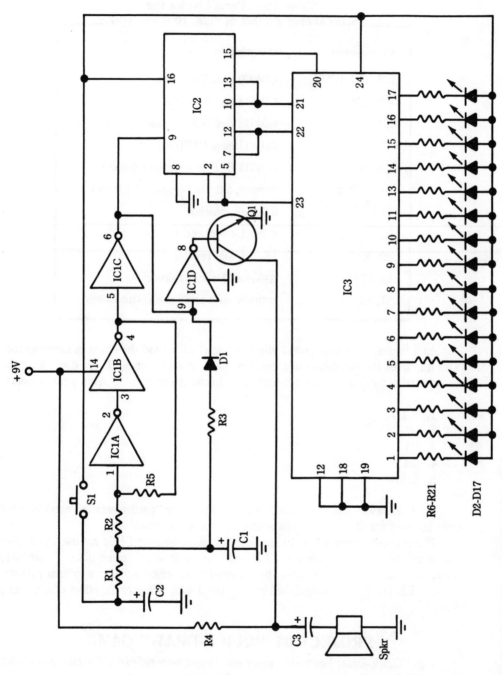

Fig. 10-8. Project 60: electronic roulette wheel.

**Table 10-5. Parts List for the
Electronic Roulette Wheel Project of Fig. 10-8.**

Component	Description
IC1	7404 hex inverter
IC2	74175 quad D-type flip-flop
IC3	74154 4-to-16 line decoder
Q1	npn transistor (2N2222, or similar)
D1	diode (1N4148, or similar)
D2–D17	LED
C1, C3	2.2-μF, 25-volt electrolytic capacitor
C2	330-μF, 25-volt electrolytic capacitor
R1	120K, 1/4-watt resistor
R2	470K, 1/4-watt resistor
R3, R4	10K, 1/4-watt resistor
R5	1 megohm, 1/4-watt resistor
R6–R21	330-ohm, 1/4-watt resistor
S1	normally open SPST pushbutton switch
Spkr	small speaker

IC1A and IC1B form a high-speed clock, feeding the counter (IC2). As the counter steps through its counting cycle, it lights up each of the eight LEDs one at a time in sequence.

The object is to hit the pushbutton switch (S1) just as LED D5 lights up. If the switch is closed with any other LED (D1–D4, or D6–D8), the display freezes for a moment. You missed. The LED you "hit" will remain lit during the delay period. If you "hit" LED D5, the display blanks out, and all eight LEDs will be dark for a few seconds.

After the delay period, the counter will start cycling again, and the circuit will be automatically ready for another round. The delay time for both these actions is determined by the values of capacitor C1 and resistor R1.

PROJECT 62: ELECTRONIC COIN FLIPPER

Some games involve yes/no choices that can be determined randomly by flipping a coin. For example, a coin-flip can be used to decide which player or team will get the first move. If you're like me, you probably drop the coin at least a third of the time. Or you don't have a coin handy when you need to flip one to make a random decision.

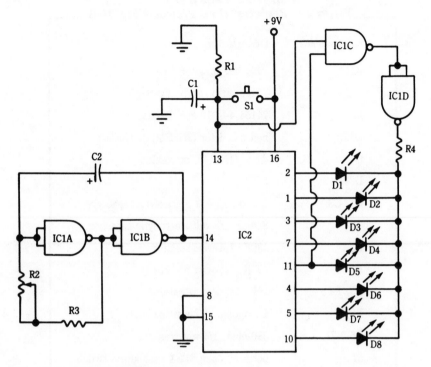

Fig. 10-9. Project 61: "quick-draw" game.

**Table 10-6. Parts List for the
"Quick Draw" Game Project of Fig. 10-9.**

Component	Description
IC1	CD4011 quad NAND gate
IC2	CD4022 counter
D1–D8	LED
C1	2.2-μF, 25-volt electrolytic capacitor
C2	1-μF, 25-volt electrolytic capacitor
R1	10K, 1/4-watt resistor
R2	100K potentiometer
R3	22K, 1/4-watt resistor
R4	390-ohm, 1/4-watt resistor
S1	normally open SPST pushbutton switch

Even if you have a coin and you flip it beautifully, it's still a rather inelegant approach in this day and age.

Figure 10-10 shows a circuit for an electronic coin flipper. A parts list for this project appears in Table 10-7.

If possible, one of the LEDs should be red and the other one should be green. It doesn't matter which is which—just so they're different from one another. You can label the red LED "Tails" and the green LED "Heads". Or, if you'd rather call this project an "Executive Decision Maker," the red LED can be labelled "No", and the green LED can be labelled "Yes." It doesn't matter how you label the two LEDs, of course. Choose something appropriate to your specific application.

Switch S1 is a common momentary action (normally open) pushbutton switch with SPST contacts. Briefly depress and release this switch to electronically "flip the coin."

This circuit is so simple there really isn't much else to say about it here.

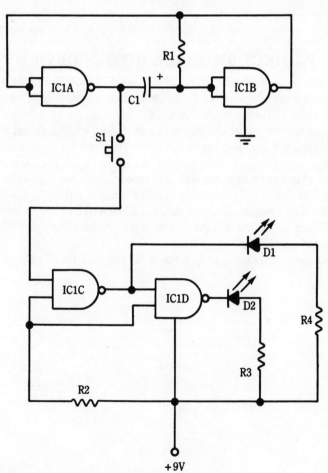

Fig. 10-10. Project 62: electronic coin flipper.

**Table 10-7. Parts List for the
Electronic Coin Flipper Project of Fig. 10-10.**

Component	Description
IC1	CD4011 quad NAND gate
D1, D2	LED
C1	2.2 μF, 25-volt electrolytic capacitor
R1	22K, 1/4-watt resistor
R2	1K, 1/4-watt resistor
R3, R4	330 ohm, 1/4-watt resistor
S1	normally open SPST pushbutton switch

PROJECT 63: QUICK RING-IN DEVICE

In many games, several competing players get a chance to answer a question or make a move, depending on which of them signals first. An example of a game that uses this approach is the television show "Jeopardy."

Determining the first player to signal is usually difficult without the aid of electronics, but the circuit shown in Fig. 10-11 makes it easy. The parts list for this project is given in Table 10-8.

Up to four players can compete with this device. Each is assigned his own switch (S1-S4). The first switch to be depressed causes the appropriate (similarly numbered) LED to light up. For example, S1 lights up D1, S2 lights up D2, and so forth. The other three switches are locked out as soon as one is depressed. Only one of the LEDs can ever be lit at any given time.

Switch S5 clears the circuit and resets it for the next round of play.

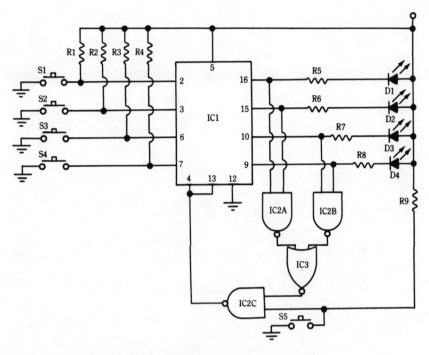

Fig. 10-11. Project 63: quick ring-in device.

Table 10-8. Parts List for the Quick Ring-In Device Project of Fig. 10-11.

Component	Description
IC1	74C75 quad latch
IC2	CD4011 quad NAND gate[*]
IC3	CD4001 quad NOR gate[*]
D1–D4	LED
R1–R4	1.2K, 1/4-watt resistor
R5–R8	390-ohm, 1/4-watt resistor
R9	1K, 1/4-watt resistor
S1–S5	normally open SPST pushbutton switch

[*]Power supply connections not shown.

Chapter 11

Communications Projects

Communicating on a beam of light is an exciting concept that still seems very futuristic, even though it has been a practical reality for years now. Many telephone services use fiber-optic cables that carry signals in the form of light.

The projects in this chapter are far less sophisticated than modern commercial systems, of course, but they are still fascinating to experiment with.

PROJECT 64: LIGHT MORSE-CODE TRANSMITTER

It would be hard to come up with a project much simpler than this one.

Morse Code is one of the simplest forms of electrical communication. The signal consists of a series of "dots" and "dashes," or short and long pulses. These pulses are arranged according to a special coding system. Each letter, numeral, and common punctuation symbol is represented by a unique pattern of pulses.

The Morse Code was first created for distant communication over telegraph wires, but it can be used in any transmission medium in which the signal can be keyed on and off according to the code patterns.

A simple light-wave Morse-Code transmitter circuit is shown in Fig. 11-1. A parts list would be superfluous for this simple project. All that is needed is a power source (a battery will do), a light generator, such as an LED (or a small lamp bulb), and a keying switch. While any normally-open momentary action switch can be used, for the most reliable operation, especially when transmitting at high speeds, a key-switch specifically designed for code transmission is recommended.

Obviously, this transmitter will work best in a darkened area where ambient light will not interfere with the coded light-beam signal. Alternatively, you could use an infrared LED as the signal source. Of course, the receiver would have to have a suitable sensor.

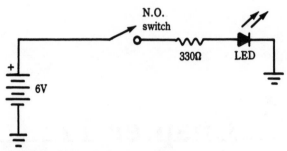

Fig. 11-1. Project 64: light Morse Code transmitter.

You could use a fiber-optic cable to connect the transmitter and the receiver. This provides the most reliable and interference-free reception, but you will lose the advantage of wireless communication.

Two suitable receivers are presented as projects 65 and 66.

PROJECT 65: LIGHT MORSE-CODE RECEIVER

To receive light-wave signals encoded with Morse Code, you need a receiver circuit, of course. Figure 11-2 shows a super-simple circuit for this purpose.

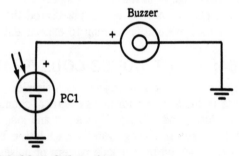

Fig. 11-2. Project 65: light Morse Code receiver.

This is one of the simplest projects in this book. In fact, it is one of the simplest projects anywhere, because it consists of just two parts. The parts list, such as it is, appears in Table 11-1.

Table 11-1. Parts List for the Light Morse Code Receiver Project of Fig. 11-2.

Component	Description
PC1	photovoltaic cell
BZ1	piezoelectric buzzer

The photocell (PC1) serves as both the signal sensor and the receiver power source. When no light is shining on the photocell, it puts out no power and nothing happens.

When a light pulse is received, the photocell puts out a small voltage (about half a volt). This is enough to cause the piezoelectric buzzer to sound. Each time a light pulse is received, a brief buzz-tone is heard. That's all there is to this project.

Of course, the sensor (PC1) must be shielded from all light except the desired encoded signal. This receiver works best in the dark.

If you'd like a visual indication of a received pulse in addition to the buzz-tone, you can modify the circuit, as shown in Fig. 11-3. Each time a light pulse is received, the buzzer will sound and the LED will glow. Mount the LED so it doesn't shine on the photocell sensor for cleaner reception. The LED's glow will not be sufficient to trigger the circuit on its own, but it can slow down the cutoff time after an incoming pulse ends. This could create minor problems if the encoded pulses are very closely spaced.

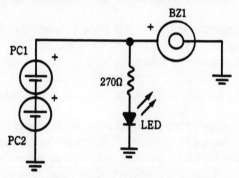

Fig. 11-3. This modification of Project 65 gives a visual as well as an audible indication of the received signal.

PROJECT 66: ALTERNATE LIGHT MORSE-CODE RECEIVER

If, for some reason, you don't particularly like project 65, I offer an alternate version here.

A second light Morse-Code receiver circuit is shown in Fig. 11-4. Table 11-2 shows the parts list.

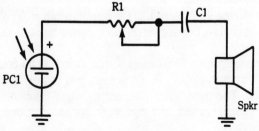

Fig. 11-4. Project 66: alternate light Morse Code receiver.

**Table 11-2. Parts List for the Alternate
Light Morse Code Receiver Project of Fig. 11-4.**

Component	Description
PC1	photovoltaic cell
Spkr	small speaker
R1	100-ohm potentiometer
C1	0.05-μF capacitor

This project isn't much more complicated than the previous one. When the photovoltaic cell (PC1) receives an incoming light pulse, a click will be heard from the speaker.

Capacitor C1 protects the speaker's coil from a constant application of dc voltage if the sensor is continuously exposed to light for some reason.

Potentiometer R1 is a simple volume control.

R1 and C1 can be eliminated in some cases, so experiment. Be careful if you do not use the protective capacitor. Eliminating the volume control (R1) should not cause you any problems.

Otherwise, this circuit performs exactly like the one in project 59. Some people might find the clicking sound produced by this circuit preferable to the buzz-tone of project 59.

This circuit can also be intriguing on a dark, stormy night for listening to flashes of lightning. It works best out in the country. City lights tend to reduce the circuit's sensitivity.

The same basic idea can easily be expanded by adding an amplifier stage to the circuit to help you catch weak signals (flashes of light).

PROJECT 67: SIMPLE TONE TRANSMITTER

The circuit shown in Fig. 11-5 is a simple optical transmitter. A suitable parts list for this project is given in Table 11-3. A suitable receiver will pick up a constant tone as long as the transmit switch (S1) is closed. The exact frequency of the transmitted signal is set by potentiometer R2.

If you prefer, you can use an infrared LED as the transmitter. The transmitted signal will then be invisible. Of course, the sensor on the receiver must be sensitive to infrared light.

Switch S1 is a normally open SPST pushbutton switch. The transmitter emits its signal only while this switch is closed. If you prefer, you can eliminate this switch and use a direct connection. The transmitter puts out its signal continuously unless power is interrupted. If the switch is used, the signal can be encoded using Morse Code or any other on/off system.

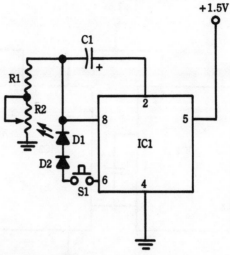

Fig. 11-5. Project 67: simple tone transmitter.

**Table 11-3. Parts List for the Simple
Tone Transmitter Project of Fig. 11-5.**

Component	Description
IC1	LM3909 flasher/oscillator
D1	LED
D2	diode (1N4148, or similar)
C1	1-μF, 10-volt electrolytic capacitor
R1	680-ohm, 1/4-watt resistor
R2	1K potentiometer

This circuit could be used as a simple remote controller. You might want to build two or three transmitters in a single case (they're very small). By using tone decoders in the receiver, various functions can be independently controlled.

PROJECT 68: TWO-TONE TRANSMITTER

This two-tone transmitter project is very similar to project 66, but it puts out a more distinctive signal. In this circuit, which is shown in Fig. 11-6, two oscillators are used. Table 11-4 shows a suitable parts list. One oscillator modulates the second, producing an unusual warbling sound.

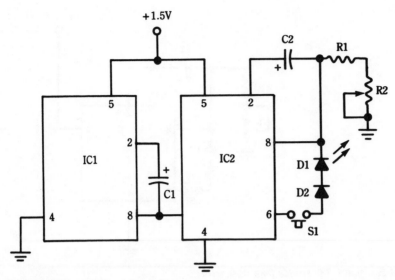

Fig. 11-6. Project 68: two-tone transmitter.

Table 11-4. Parts List for the Two-Tone Transmitter Project of Fig. 11-6.

Component	Description
IC1, IC2	LM3909 flasher/oscillator
D1	LED
D2	diode (1N4148, or similar)
C1	50-μF, 10-volt electrolytic capacitor
C2	1-μF, 10-volt electrolytic capacitor
R1	180-ohm, 1/4-watt resistor
R2	2.5K potentiometer

PROJECT 69: OPTICAL VOICE TRANSMITTER

In more advanced communications systems, we want to be able to transmit audio signals, such as a speaking voice. A very simple optical-voice-transmitter circuit is illustrated in Fig. 11-7. Table 11-5 shows the parts list for this project.

The input is from a dynamic microphone, or a similar signal source.

Potentiometer R6 is a distortion control. Adjust this potentiometer for the clearest possible signal. The setting is usually fairly critical, so take your time and work carefully. Because this is a calibration control, I advise using a trimpot for R6.

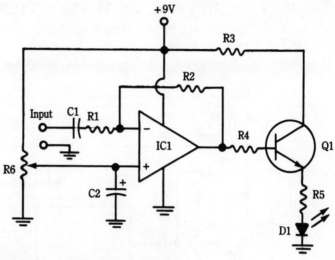

Fig. 11-7. Project 69: optical voice transmitter.

**Table 11-5. Parts List for the Optical
Voice Transmitter Project of Fig. 11-7.**

Component	Description
IC1	op amp
Q1	npn transistor (2N3904, or similar)
D1	LED
C1	0.33-μF capacitor
C2	10-μF, 25-volt electrolytic capacitor
R1	1.8K, 1/4-watt resistor
R2	180K, 1/4-watt resistor
R3, R4, R5	100-ohm, 1/4-watt resistor
R6	1-megohm potentiometer

The transmitter (D1) can be a visible LED or an infrared LED. Remember that all optical communication systems are line-of-sight only. Light signals cannot go around corners or through objects. In some applications, it might be desirable to connect the transmitter to the receiver with a fiber-optic cable (see chapter 4).

PROJECT 70: OPTICAL FM TRANSMITTER

This project is the most elaborate optical transmitter in this chapter. It uses frequency modulation (FM) to carry the voice signal.

Figure 11-8 shows the circuit diagram, and the parts list appears in Table 11-6.

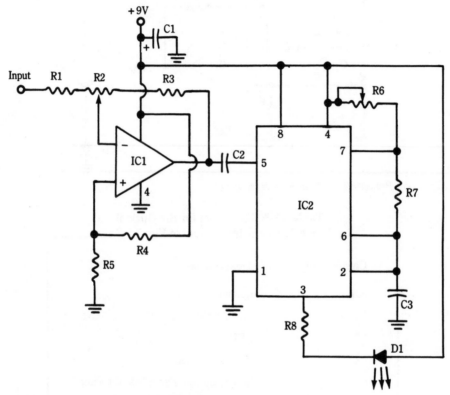

Fig. 11-8. Project 70: optical FM transmitter.

The input signal is taken from a microphone or a similar audio signal source.

Potentiometer R6 is a calibration control. Adjust this potentiometer for the clearest possible signal at the receiver. It would make sense to use a trimpot for R6. This potentiometer is actually controlling the frequency of the FM carrier. The range is about 35 to 45 kHz.

The other potentiometer in this circuit is a volume or signal-level control. In some applications, this is a front-panel control, while in other applications it will be a "set and forget" calibration trimpot.

The transmitter (D1) can be a visible LED or an infrared LED. Remember that all optical communications systems are line-of-sight only. Light signals cannot go around corners or through objects. With suitable lenses, however, this transmitter can send

**Table 11-6. Parts List for the
Optical FM Transmitter Project of Fig. 11-8.**

Component	Description
IC1	op amp
IC2	555 timer
D1	LED
C1	10-μF, 25-volt electrolytic capacitor
C2	0.1-μF capacitor
C3	500-pF capacitor
R1, R3	1K, 1/4-watt resistor
R2	100K potentiometer
R4, R5	5.6K, 1/4-watt resistor
R6	100K potentiometer
R7	12K, 1/4-watt resistor
R8	220-ohm, 1/4-watt resistor

signals hundreds of feet in a line-of-sight path. In some applications, it might be desirable to connect the transmitter to the receiver with a fiber-optic cable (see chapter 4).

A suitable FM receiver for this transmitter project is discussed later in this chapter in project 73.

PROJECT 71: FIRST OPTICAL RECEIVER

One of the simplest possible optical receiver circuits is shown in Fig. 11-9. Table 11-7 shows the parts list. This receiver can be used with any of the transmitters of projects 67, 68, or 69.

If the transmitter uses an infrared LED as the signal emitter, the sensor on the receiver (phototransistor Q1) should be selected to be sensitive to infrared rather than visible light.

Almost any op-amp chip can be used for IC1, but a high-grade device, such as the TL084, is strongly recommended for this type of application. The output of this circuit can be fed into any standard audio amplifier.

PROJECT 72: SECOND OPTICAL RECEIVER

The circuit illustrated in Fig. 11-10 is quite similar to project 70. The main difference is that this optical receiver circuit features a built-in speaker. No external amplifier is required.

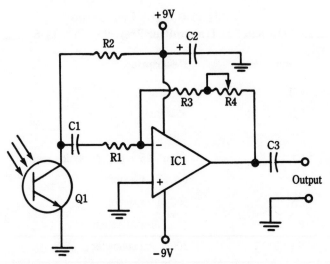

Fig. 11-9. Project 71: first optical receiver.

**Table 11-7. Parts List for the First
Optical Receiver Project of Fig. 11-9.**

Component	Description
IC1	op amp (TL084, or similar)
Q1	phototransistor
C1, C3	0.1-μF capacitor
C2	10-μF, 25-volt electrolytic capacitor
R1	2.2K, 1/4-watt resistor
R2	220K, 1/4-watt resistor
R3	1K, 1/4-watt resistor
R4	500K potentiometer

Potentiometer R3 serves as a volume control for the receiver.
A suitable parts list for this project is given in Table 11-8.

PROJECT 73: THIRD OPTICAL RECEIVER

This third optical receiver project is slightly more sophisticated than project 70 or project 71. Figure 11-11 shows the circuit diagram for this project. Table 11-9 is the parts list.

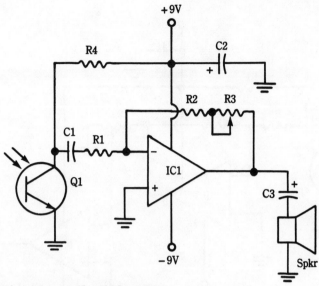

Fig. 11-10. Project 72: second optical receiver.

**Table 11-8. Parts List for the Second
Optical Receiver Project of Fig. 11-10.**

Component	Description
IC1	op amp
Q1	phototransistor
C1	0.1-μF capacitor
C2	10-μF, 25-volt electrolytic capacitor
C3	5-μF, 25-volt electrolytic capacitor
R1	1.2K, 1/4-watt resistor
R2	100K, 1/4-watt resistor
R3	500K potentiometer
R4	220K, 1/4-watt resistor
Spkr	small 8-ohm speaker

Because an audio amplifier stage (IC2) is built-in, this circuit features higher gain and can reproduce weaker transmitted signals. The amplifier's gain is fixed at 20. The output volume from the speaker is controlled by adjusting the signal level at the input of the final amplification stage.

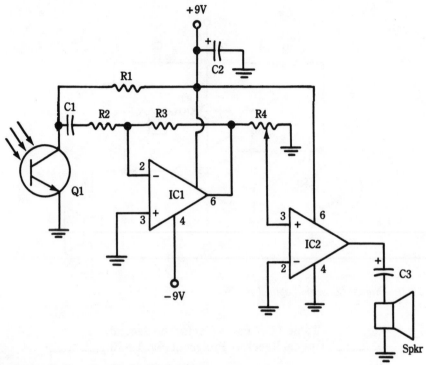

Fig. 11-11. Project 73: third optical receiver.

**Table 11-9. Parts List for the Third
Optical Receiver Project of Fig. 11-11.**

Component	Description
IC1	op amp (TL084, or similar)
IC2	LM386 audio amplifier
Q1	phototransistor
C1	0.1-μF capacitor
C2	10-μF, 25-volt electrolytic capacitor
C3	100-μF, 25-volt electrolytic capacitor
R1	220K, 1/4-watt resistor
R2	1.2K, 1/4-watt resistor
R3	12K, 1/4-watt resistor
R4	25K potentiometer
Spkr	small 8-ohm speaker

Potentiometer R4 is a volume control for the receiver as a whole. Be careful when using this receiver. Always start out with the volume control at its lowest setting and slowly turn it up to a comfortable listening level. A gain of 20 is a pretty significant amount of amplification. This receiver can put out some very loud sounds.

Almost any op amp can be used as IC1, but I strongly recommend a high-grade device, such as the TL084, for this project.

This receiver is extremely sensitive, so the transmitter can be placed a little further away. The system's range can be improved even further by using lenses for both the transmitter and the receiver, and/or connecting the transmitter and the receiver with a fiber-optic cable (see chapter 4).

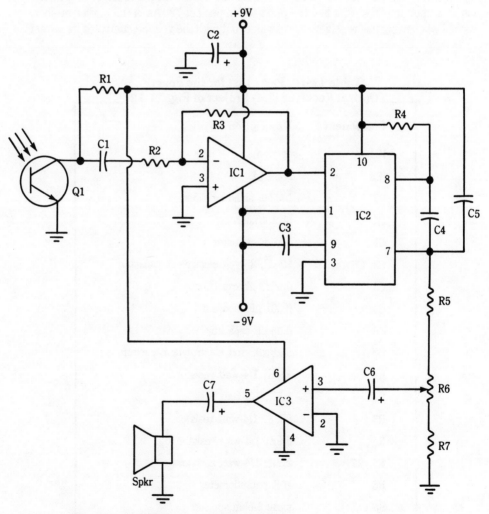

Fig. 11-12. Project 74: fourth optical receiver (FM).

PROJECT 74: FOURTH OPTICAL RECEIVER (FM)

This final receiver project is intended to be used with the optical FM transmitter of project 69. The circuit is illustrated in Fig. 11-12, with the parts list appearing as Table 11-10.

Almost any op-amp chip may be used for IC1. For high-fidelity communications, I recommend a high-grade op-amp device, such as the TL084.

For the component values given in the parts list, the receiver is tuned to a frequency of about 40.5 kHz. To tune the receiver to a different frequency, change the value of capacitor C3.

The received FM signal is decoded by a phase-locked loop (PLL) (IC2 and its related components). Potentiometer R6 serves as the volume control for the receiver as a whole. IC3 is an output amplifier, just like the one used in project 72. As in the earlier project, be careful when using this receiver. Always start out with the volume control at its lowest

Table 11-10. Parts List for the Fourth Optical Receiver (FM) Project of Fig. 11-12.

Component	Description
IC1	op amp
IC2	565 PLL
IC3	LM386 audio amplifier
Q1	phototransistor
C1	0.1-μF capacitor
C2, C6	10-μF, 25-volt electrolytic capacitor
C3	0.0022-μF capacitor
C4	0.001-μF capacitor
C5	0.05-μF capacitor
C7	5-μF, 25-volt electrolytic capacitor
R1	220K, 1/4-watt resistor
R2	1.5K, 1/4-watt resistor
R3	680K, 1/4-watt resistor
R4	3.9K, 1/4-watt resistor
R5, R7	2.2K, 1/4-watt resistor
R6	10K potentiometer
Spkr	small 8-ohm speaker

setting and slowly turn it up to a comfortable listening level. A gain of 20 is a pretty significant amount of amplification. This receiver can produce very loud sounds.

For best results with an optical FM communications system, a fiber-optic cable interconnecting the transmitter (project 69) and the receiver (project 73) is strongly advised.

If the receiver (or the transmitter) exhibits an oscillation problem, small filter capacitors (about 0.1 μF) across the power supply connections should help. Keep the power supply leads as short as possible.

Chapter 12

Photography and Light-Meter Projects

This chapter is almost obligatory for this book. Electronic circuits for use with photography, either on the hobbyist or the professional level, are a natural application for optoelectric devices. After all, the word *photography* breaks down into "light writing."

Many of the projects in this chapter are light meters of various types. Obviously, a photographer needs to know as much as possible about lighting levels, both general ambient lighting, and light intensity at specific points.

PROJECT 75: SIMPLE LIGHT METER

If your photography work is a little more demanding, you will need a light meter that does more than give a simple "yes" or "no" indication. Probably the simplest possible circuit is shown in Fig. 12-1. The parts list for this project is given in Table 12-1. Note that only three components are required for this simple project.

M1 is the actual meter. This is a simple dc milliammeter that can read from 0 to 1 mA. You might want to design a specially calibrated light-meter scale to replace the current scale supplied with the meter. Be very careful when opening the meter's case. There are some very small parts that can easily be lost or damaged. Do not force anything.

If you don't want to open up the meter's housing, you can make up a simple conversion chart and mount it on the side of the project for easy reference.

Potentiometer R1 is a sensitivity control. If your application is very non-critical, you could substitute a fixed resistor for this potentiometer. If you do use the sensitivity control as shown here (advised for most applications), you will need to decide whether you want this to be a front panel control, or a "set and forget" calibration control. In the second case, use a screwdriver-turned trimpot for R1.

R2 is the actual light sensor for this project. It is just a common photoresistor. Any type will do.

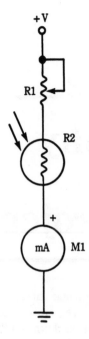

Fig. 12-1. Project 75: simple light meter.

**Table 12-1. Parts List for the
Simple Light-Meter Project of Fig. 12-1.**

Component	Description
R1	100K potentiometer or trimpot
R2	photoresistor (cadmium sulfide)
M1	dc milliammeter (0–1 mA range)

The supply voltage for this project is not critical. It works fine over a wide range of voltages. Of course, the voltage used affects the calibration of the meter. For most applications, a standard 9-volt "transistor" battery is probably the most convenient choice.

This simple light meter can operate over a very wide range. For best accuracy, the sensor should be mounted so that it can measure the light at a specific point without excessive interference from light from other areas. You might want to place some kind of shading shield around the sensor in some applications.

PROJECT 76: IMPROVED LIGHT METER

This project is a more sophisticated light meter than Project 75. A photovoltaic cell is used as the sensor. Its current output is amplified by an FET to drive a milliammeter.

The circuit is shown in Fig. 12-2. The parts list for this project appears in Table 12-2.

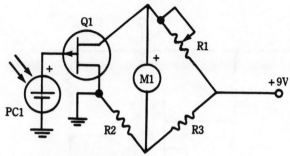

Fig. 12-2. Project 76: improved light meter.

**Table 12-2. Parts List for the
Improved Light-Meter Project of Fig. 12-2.**

Component	Description
Q1	FET (2N3578, or similar)
PC1	photovoltaic cell
M1	milliammeter (0–1 mA dc)
R1	5K potentiometer
R2, R3	470-ohm, 1/2-watt resistor

The drain-to-source resistance of the FET (Q1) is one leg of a bridge circuit. The other four legs of the bridge are made up of R1, R2, and R3. Potentiometer R1 is the calibration control to adjust the meter for a true zero reading when the sensor (PC1) is dark (shaded from external light).

The use of the FET amplifier in this circuit increases the photocell's sensitivity by a factor of two or three. The exact increase in sensitivity depends on the specific characteristics of the individual photovoltaic cell and FET you use in your project. With any functional combination, there is a significant increase in sensitivity.

PROJECT 77: LIGHT-RANGE DETECTOR

In some applications, you don't really need to know the exact light intensity. You are only concerned with whether or not the light intensity falls within a specified acceptable range of values. The circuit shown in Fig. 12-3 will tell you what you need to know. The parts list for this project is given in Table 12-3.

Basically, this circuit is just a window comparator. Two voltage-comparator stages are used. The upper end of the acceptable range, or "window," is controlled by IC1A, while IC1B sets the lower end of the acceptable range.

The reference voltages for the two comparator stages are set by the simple resistive voltage-divider network made up of resistors R1, R2, and R3. Experiment with other

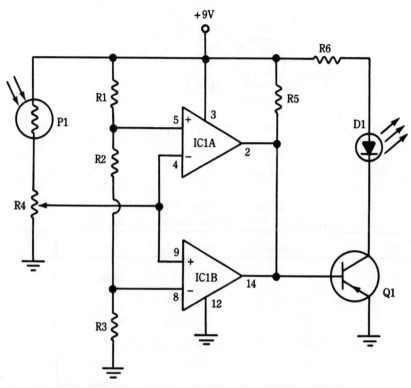

Fig. 12-3. Project 77: light-range detector.

**Table 12-3. Parts List for the
Light-Range Detector Project of Fig. 12-3.**

Component	Description
IC1	LM339 quad comparator
Q1	pnp transistor (2N3906, or similar)
D1	LED
P1	photoresistor
R1, R3	100K, 1/4-watt resistor
R2	220K, 1/4-watt resistor
R4	500K potentiometer
R5	10K, 1/4-watt resistor
R6	330-ohm, 1/4-watt resistor

values for these three resistors. The three resistance values inevitably interact, but you can roughly divide them into separate functions, as follows:

R1   sets upper value
R2   sets range width
R3   sets lower value

The larger R2's resistance is, the wider the acceptable range or "window" will be.

The input voltage to be compared with the reference voltages is also set by a resistive voltage-divider network. This time the network is made up of photoresistor (P1) and potentiometer R4. This potentiometer acts as a sensitivity control for the sensor. You could substitute a fixed resistor for R4 if your individual application does not require a manual sensitivity control. It is best to breadboard the circuit and determine the desired value of R4 experimentally.

The amount of light striking the photoresistor determines its resistance. The lower this resistance is, the higher the resulting input voltage, and vice versa.

If the light intensity detected by the photoresistor is within the acceptable range, the circuit's output goes high, and the LED will light up. In some applications, you might want to eliminate the LED and use the comparator output to drive some other circuit. A buffer amplifier might be required in some cases.

PROJECT 78: ALTERNATE LIGHT-RANGE METER

This project is a variation on the light-range meter of Project 77. The circuit diagram appears in Fig. 12-4, and Table 12-4 gives the parts list.

Potentiometer R1 sets the sensitivity of the sensor (photoresistor R1). Potentiometer R3 sets the range. If the light striking the sensor is within the range (determined by the settings of potentiometers R1 and R3), the output LED (D1) lights up, otherwise, the LED remains dark.

The two potentiometers can be manual front-panel controls, or they can be screwdriver-adjusted calibration trimpots. The choice depends on your individual application. Will you need to reset the settings for different conditions, or do you need one constant set of adjustments that shouldn't be changed?

Because the circuit uses just two sections (one half) of a LM339 quad comparator IC, you might want to consider building two light-range meters in a single housing. Each meter will be adjusted for a different, but overlapping range. The two output LEDs in tandem will give you more information about the actual light level.

PROJECT 79: ALTERNATE SIMPLE LIGHT METER

This light meter project takes a completely different tack than any of the other circuits presented in this chapter. The circuit is shown in Fig. 12-5. The parts list for this project appears in Table 12-5.

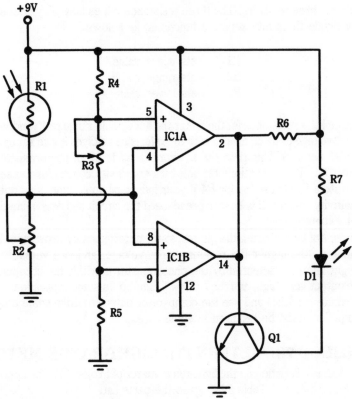

Fig. 12-4. Project 78: alternate light-range meter.

Table 12-4. Parts List for the Alternate Light-Range Meter Project of Fig. 12-4.

Component	Description
IC1	LM339 quad comparator
Q1	npn transistor (2N3904, 2N2222, or similar)
D1	LED
R1	photoresistor
R2	1-megohm potentiometer
R3	500K potentiometer
R4, R5	47K, 1/4-watt resistor
R6	10K, 1/4-watt resistor
R7	470 ohm, 1/4-watt resistor

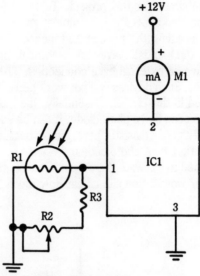

Fig. 12-5. Project 79: alternate simple light meter.

**Table 12-5. Parts List for the Alternate,
Simple Light-Meter Project of Fig. 12-5.**

Component	Description
IC1	LM334 temperature sensor/adjustable current source
R1	photoresistor
R2	500K potentiometer
R3	470K, 1/4-watt resistor
M1	milliammeter (0- to 1-mA scale)

The heart of this circuit is the LM334 IC. This is a dual-function device. It is marketed as a temperature sensor and as an adjustable current source. We are using the LM334 as a constant-current source in this project.

The relative light level is indicated on the milliammeter (M1). Photoresistor R1 is the light sensor for this simple light meter.

Potentiometer R2 has a range control, adjusting the sensitivity of the photoresistor. In some applications, R2 and R3 can be eliminated.

PROJECT 80: LIGHT METER WITH BARGRAPH DISPLAY

We worked with bargraph displays in several of the earlier projects in this book. This project is a light meter with a dedicated five-step bargraph to display the lighting level. This output display is very easy to read at a glance.

Figure 12-6 shows the circuit for this project. Table 12-6 shows the parts list.

This is not a critical op-amp application, so almost any op-amp IC can be used in constructing this project, including the popular and inexpensive 741.

The phototransistor (Q1) must have an electrical base connection. Many phototransistors do not have an external base connection. The light source serves as the only base current source. Such devices will not work here. The emitter, on the other hand, is left unconnected in this circuit. Essentially, the base-connector junction is being used as a diode. If you can find a photodiode, it can be substituted in place of the phototransistor (Q1). I used a phototransistor here, simply because they tend to be easier to find on the hobbyist market than photodiodes.

Potentiometer R1 is used to calibrate the output. This control allows you to set the zero reading level. In other words, you can adjust how much light is required to turn on the first LED.

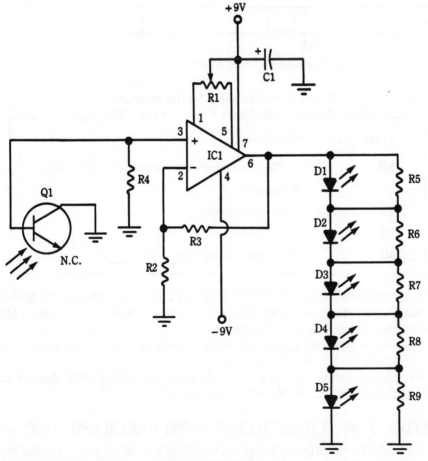

Fig. 12-6. Project 80: light meter with bargraph display.

**Table 12-6. Parts List for the Light-Meter-
with-Bargraph-Display Project of Fig. 12-6.**

Component	Description
IC1	op amp (741, or similar)
Q1	phototransistor
D1–D5	LED
C1	10-μF, 25-volt electrolytic capacitor
R1	10K potentiometer
R2	680K, 1/4-watt resistor
R3	68K, 1/4-watt resistor
R4	1K, 1/4-watt resistor
R5	150-ohm, 1/4-watt resistor
R6	220-ohm, 1/4-watt resistor
R7	330-ohm, 1/4-watt resistor
R8	390-ohm, 1/4-watt resistor
R9	470-ohm, 1/4-watt resistor

You might want to experiment with other values for resistors R5 through R9, although these particular values given in the parts list seem to work about the best.

As with most op-amp circuits, a dual-polarity power supply is required for this project.

PROJECT 81: LOGARITHMIC-READING LIGHT METER

All of the light-meter projects presented earlier in this chapter function along a linear scale. In some applications, we may need a light meter that reads out in logarithmic fashion. A suitable circuit for this appears in Fig. 12-7. The parts list for this project is given in Table 12-7.

The "secret" of this circuit is that the photosensor is a diode (D1). The nonlinear current/voltage relationship of the pn junction causes the milliammeter (M1) to move logarithmically rather than linearly in response to the lighting level. If you cannot find a photodiode, you could use the base–collector junction of a phototransistor as a diode. Of course, the phototransistor must have an external base lead to be used for this application.

The sensitivity of this logarithmic light meter can be controlled by potentiometer R3. In some applications, this can be a "set and forget" calibration trimpot.

At very low lighting levels, the output of the op amp can go negative. This could peg the meter pointer and damage it. To prevent this possibility, diode D2 is placed in the op-amp's output path.

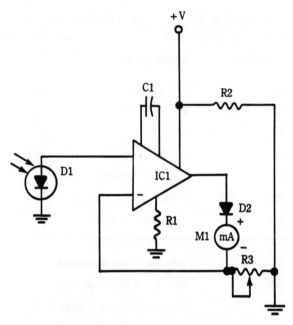

Fig. 12-7. Project 81: logarithmic-reading light meter.

Table 12-7. Parts List for the Logarithmic-Reading Light-Meter Project of Fig. 12-7.

Component	Description
IC1	op amp (LM308, or similar)
D1	photodiode
D2	diode (1N4148)
C1	30-pF capacitor
R1	5.6K, 1/4-watt resistor
R2	10K, 1/4-watt resistor
R3	10K potentiometer
M1	dc milliammeter

PROJECT 82: LIGHT-TRIGGERED REMOTE FLASH DRIVER

Any serious indoor photography beyond the simplest snapshot level usually requires multiple flash units. Getting several flash units to operate in synchrony can sometimes be a little tricky. The next few projects will help make this task a bit easier.

Fig. 12-8. Project 82: light-triggered remote flash driver.

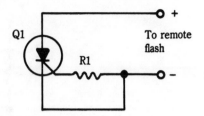

Q1

R1

To remote flash

+

−

Table 12-8. Parts List for the Light-Triggered Remote-Flash-Driver Project of Fig. 12-8.

Component	Description
Q1	LASCR (rated for at least 300 volts)
R1	47K, 1/2-watt resistor

First, this simple two-part project automatically senses the light from the main flash unit and triggers a secondary remote flash unit. Figure 12-8 shows the circuit. The parts list, short as it is, appears in Table 12-8.

Q1 is a LASCR, or light-activated, silicon-controlled rectifier. The gate is triggered by the light energy striking the photosensitive surface of the LASCR. In operation, the outputs of this circuit are connected to the remote flash unit, and the LASCR is positioned so that it faces the maximum possible exposure to the light from the main flash unit.

PROJECT 83: SECOND LIGHT-TRIGGERED REMOTE FLASH DRIVER

This project does pretty much the same thing as Project 82. Sometimes LASCRs can be a little difficult to find, and rather expensive when you do find them. In this circuit, we use a regular (non-light sensitive) SCR (Q1) to trigger the remote flash unit.

This circuit diagram is shown in Fig. 12-9. The parts list for this project appears as Table 12-9.

A photoresistor, (R1) serves as the light sensor in this remote flash driver circuit. The photoresistor is part of a resistive voltage-divider network made up of R1, R2, and R3. As the light level increases, the resistance of the photoresistor drops. At some point, the voltage at the junction of R1 and R2 is sufficient to trigger the gate of Q1, causing the SCR to fire the remote flash unit.

In operation, the outputs of this circuit are connected to the remote flash unit, and the photoresistor is positioned so that it faces the maximum possible exposure to the light from the main flash unit.

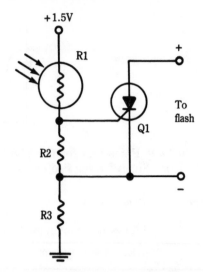

Fig. 12-9. Project 83: second light-triggered remote flash driver.

Table 12-9. Parts List for the Second Light-
Triggered Remote-Flash-Driver Project of Fig. 12-9.

Component	Description
Q1	SCR (suitable to drive flash unit) (C106, or similar)
R1	photoresistor
R2	22K, 1/2-watt resistor
R3	47-ohm, 1/2-watt resistor

PROJECT 84: SOUND-TRIGGERED FLASH

For some special photographic effects, it is handy to have a flash unit that can be triggered by a loud, sharp sound. For example, you could take a photograph of a balloon bursting, or a bullet in mid-flight as it has been fired from a gun.

A sound activated flash driver circuit is illustrated in Fig. 12-10. Table 12-10 gives the parts list for this project.

This project has a considerably higher parts count than most of the other projects in this chapter, but it isn't really a very complicated circuit. No exotic, hard-to-find parts are required.

A microphone (or other audio signal source) is connected to jack J1. In some cases, you might want to eliminate the jack and hard-wire a dedicated microphone directly to the input of this circuit.

The circuit can be adjusted via potentiometer R6 to fire the flash after a delay of 5 to 200 milliseconds (0.005 to 0.2 second).

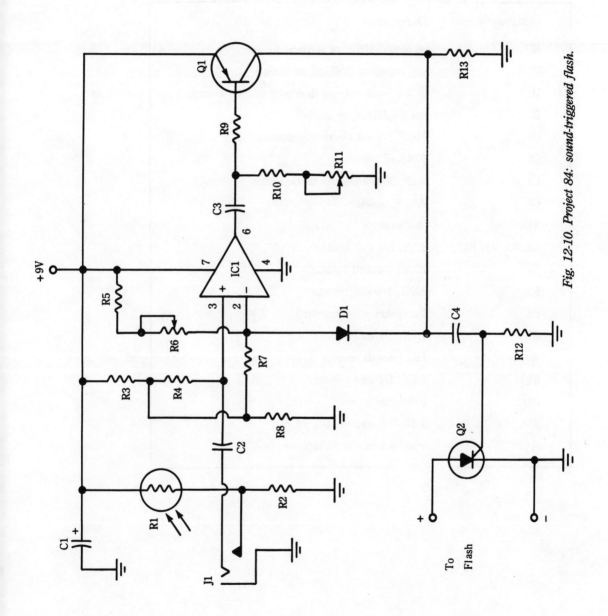

Fig. 12-10. Project 84: sound-triggered flash.

Table 12-10. Parts List for the Sound-Triggered-Flash Project of Fig. 12-10.

Component	Description
IC1	op amp (LM301, or similar)
Q1	pnp transistor (2N3905, or similar)
Q2	SCR (to suit external flash unit) (C106, or similar)
D1	diode (1N4148, or similar)
C1	10-μF, 25-volt electrolytic capacitor
C2	0.005-μF capacitor
C3	5-μF, 25-volt electrolytic capacitor
C4	0.1-μF capacitor
R1	photoresistor
R2, R3, R8, R13	3.3K, 1/4-watt resistor
R4	100K, 1/4-watt resistor
R5	220K, 1/4-watt resistor
R6	2-megohm potentiometer
R7	4.7K, 1/4-watt resistor
R9	1K, 1/4-watt resistor
R10	3.9K, 1/4-watt resistor
R11	50K potentiometer
R12	3.3K, 1/2-watt resistor
J1	input jack to suit microphone

Chapter 13
Counter Circuits

This chapter features projects using digital circuits for their most obvious and direct application—counting. For most of these projects, the optoelectric element is in the display of the counted value. Also included in this chapter is a photoelectric counter project that uses a photosensor to "see" and count actual objects.

PROJECT 85: BINARY COUNTER DEMONSTRATOR

We are used to working with the decimal number system, which has ten digits:

0, 1, 2, 3, 4,
5, 6, 7, 8, 9

The binary number system, on the other hand, has only two digits:

0, 1

Digital electronic circuits use the binary number system, instead of the more familiar decimal system. Circuit designers don't use the binary system just to be perverse. It is easier to electronically represent a two-digit system. Generally, a low voltage is used as a 0, and a relatively high voltage is used as a 1.

If you work with digital circuitry at all, it would be very helpful for you to be familiar with the binary numbering system. This binary-counter demonstrator project is a good way to get a feel for how binary numbers work. The circuit is shown in Fig. 13-1. Table 13-1 gives the parts list for this project.

IC1, along with its associated components, is a simple square-wave generator. It serves as the system clock for this project. An external signal source can be substituted, if you prefer.

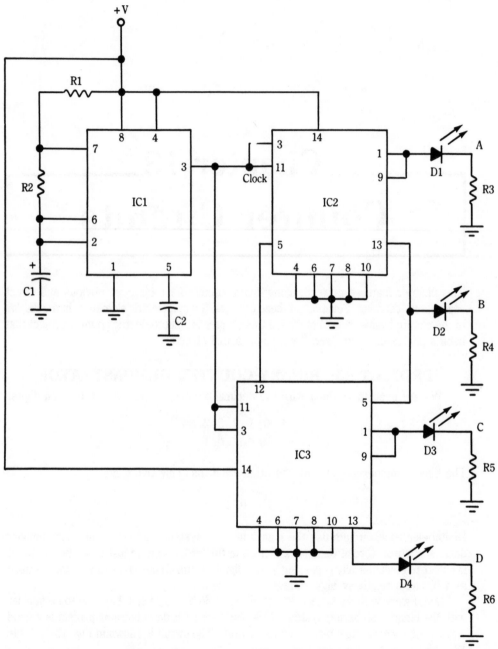

Fig. 13-1. Project 85: binary counter demonstrator.

Table 13-1. Parts List for the Binary-Counter-Demonstrator Project of Fig. 13-1.

Component	Description
IC1	555 timer
IC2, IC3	CD4013 dual D-type flip-flop
D1–D4	LED
C1	1-μF, 25-volt electrolytic capacitor
C2	0.01-μF capacitor
R1	330K, 1/4-watt resistor
R2	2.2K, 1/4-watt resistor
R3–R6	330-ohm, 1/4-watt resistor

IC2 and IC3 make up the actual counter. These chips are CD4013 dual D-type flip-flops. Because each IC contains two flip-flops, this counter has four stages. Four D-type flip-flops are connected in series.

The flip-flops count the number of clock pulses in binary. The LEDs indicate the current count value. Because there are four stages, there are sixteen (four to the second power) possible count values. Remember, zero is a step in the counting sequence, so the maximum count value is fifteen, or in binary form, 1111. Once this maximum count has been exceeded, the counter resets itself to zero (0000) and starts over.

When you watch the LEDs, they will light up in the following sequence. A binary 0 is indicated by a dark LED, while a binary 1 is represented by a lit LED.

Binary Count d c b a	Decimal Count
0 0 0 0	0
0 0 0 1	1
0 0 1 0	2
0 0 1 1	3
0 1 0 0	4
0 1 0 1	5
0 1 1 0	6
0 1 1 1	7
1 0 0 0	8
1 0 0 1	9
1 0 1 0	10
1 0 1 1	11
1 1 0 0	12
1 1 0 1	13
1 1 1 0	14

Binary Count	Decimal Count
1 1 1 1	15
0 0 0 0	0 (16)

(NOTE: the counter resets at this point)

Binary Count	Decimal Count
0 0 0 1	1 (17)
0 0 1 0	2 (18)
0 0 1 1	3 (19)
0 1 0 0	4 (20)
0 1 0 1	5 (21)

and so forth.

This pattern repeats continuously, advancing the count for each clock pulse.

PROJECT 86: SINGLE-CYCLE, VARIABLE-LENGTH COUNTER

The CD4017 decade counter is a useful chip for many applications. It has ten separate outputs. Only one output is active at any given time. The CD4017 can operate in either of two basic modes:

Count to n and halt

Count to n and recycle

In the first mode, ("count to n and halt"), the counting sequence stops when the maximum count value is reached. The other mode ("count to n and recycle") is used for redundant counting. When the maximum count is reached, the counter jumps back to 0 and then starts over. This can continue indefinitely.

In either mode, the length of the count sequence can be determined by the user. Any number of steps, up to ten, can be selected.

Figure 13-2 shows a circuit for using the CD4017 in the count-to-n-and-halt mode. Table 13-2 shows the parts list. This counter circuit counts through a single count cycle of user-determined length. Connect pin #13 to the appropriate output pin for the desired maximum count value:

Pin#	Maximum Count
3	0
2	1
4	2
7	3
10	4
1	5
5	6
6	7
9	8
11	9

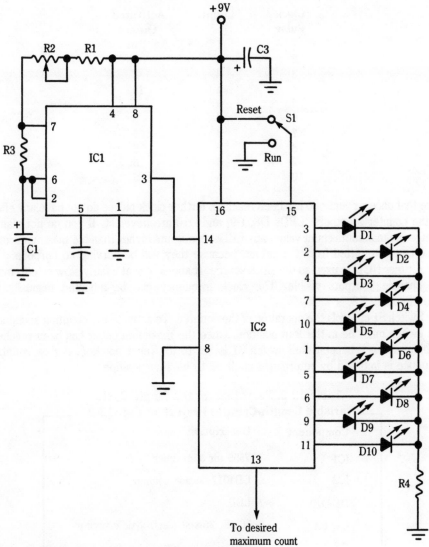

Fig. 13-2. Project 86: single-cycle variable-length counter.

For example, let's assume that pin #13 is tied to pin #5. This gives us a maximum count value of 6. The counter activates its output in this order:

Clock Pulse	Count	Activated Output Pin
0	0	3
1	1	2

Clock Pulse	Count	Activated Output Pin
2	2	4
3	3	7
4	4	10
5	5	1
6	6	5
7	6	5
8	6	5
9	6	5
10	6	5

Note that once a count of 6 has been reached, further clock pulses do not have any effect on the counter's outputs. LEDs D8, D9, and D10 are never lit. If you do not plan to change the maximum count value, any LEDs on outputs higher than the maximum count value can be omitted from the circuit, because they will be serving no purpose.

A timer (IC1) is used as the clock source. It should run at a fairly slow rate to make the changing outputs visible. The clock frequency can be adjusted manually via potentiometer R2.

Switch S1 controls the operation of the counter. To activate the counting sequence, this switch must be in the Run position. Once the maximum count has been reached, nothing else will happen until switch S1 is set to its Reset position. A new counting sequence is initiated by returning switch S1 to its Run position.

Table 13-2. Parts List for the Single-Cycle, Variable Length-Counter Project of Fig. 13-2.

Component	Description
IC1	7555 (or 555) timer
IC2	CD4017 decade counter
D1–D10	LED
C1, C3	10-μF, 25-volt electrolytic capacitor
C2	0.01-μF capacitor
R1, R3	10K, 1/4-watt resistor
R2	500K potentiometer
R4	470-ohm, 1/4-watt resistor
S1	SPDT switch

PROJECT 87: REPEATING-CYCLE, VARIABLE-LENGTH COUNTER

This project is very similar to Project 86, except this time the CD4017 decade counter is being used in its count-to-n-and recycle mode.

The circuit for this repeating cycle, variable-length counter project is illustrated in Fig. 13-3. Table 13-3 gives the parts list.

This circuit is basically the same as the one in Fig. 13-2. The biggest difference here is that pin #15 (rather than pin #13) is connected to the appropriate output pin to set the desired maximum count value. Pin #13 is grounded in this circuit.

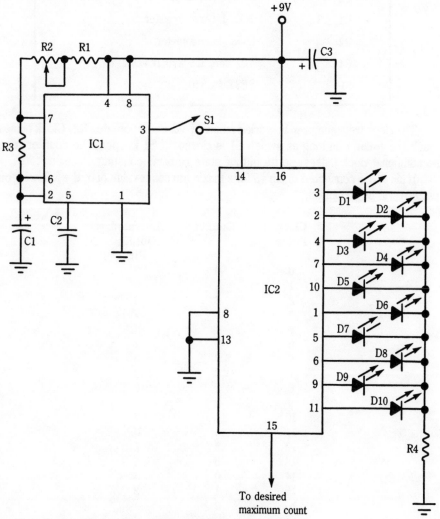

Fig. 13-3. Project 87: repeating-cycle, variable-length counter.

Table 13-3. Parts List for the Repeating-Cycle, Variable-Length-Counter Project of Fig. 13-3.

Component	Description
IC1	7555 (or 555) timer
IC2	CD4017 decade counter
D1–D10	LED
C1, C3	10-μF, 25-volt electrolytic capacitor
C2	0.01-μF capacitor
R1, R3	10K, 1/4-watt resistor
R2	500K potentiometer
R4	470-ohm, 1/4-watt resistor
S1	SPST switch

The clock frequency can be varied manually via potentiometer R2. Clock pulses can reach the counter as long as switch S1 is closed. If S1 is open, the counter receives no additional clock pulses, so its output state remains constant.

If pin #15 is connected to pin #5 for a maximum count value of 6, the following output pattern will be exhibited:

Clock Pulse	Count	Activated Output Pin
0	0	3
1	1	2
2	2	4
3	3	7
4	4	10
5	5	1
6	6	5
7	0	3
8	1	2
9	2	4
10	3	7
11	4	10
12	5	1
13	6	5
14	0	3
15	1	2
16	2	4

and so forth. This pattern continues indefinitely, as long as switch S1 is closed and power is applied to the circuit. Note that 0 is a count step. The full sequence has seven steps to reach a maximum value of 6.

As in Project 86, outputs above the maximum count value are never activated. If the maximum count value will not be changed, the LEDs for these higher valued, permanently inactive outputs (D8, D9, and D10, in our example) can be omitted, as they serve no purpose.

PROJECT 88: EXTENDED-RANGE, DUAL-DECADE COUNTER

The CD4017 decade counter (used in Projects 86 and 87) is a very useful device, but its maximum count range has just ten possible steps. Many applications require a larger counter range than this.

Two (or more) CD4017 counter ICs can be used together to create an extended range counter. Figure 13-4 shows a dual-decade counter that can count up to 100 steps (from 00 to 99). The parts list for this project is given in Table 13-4.

Just twenty LEDs are used to indicate all possible values from 00 to 99. Two LEDs are indicated for any given count. One set of LEDs (D1–D10) indicates the units column of a two-digit number:

D1	0
D2	1
D3	2
D4	3
D5	4
D6	5
D7	6
D8	7
D9	8
D10	9

The second set of LEDs (D11–D12) indicates the tens column of the two digit count value:

D11	00
D12	10
D13	20
D14	30
D15	40
D16	50
D17	60
D18	70
D19	80
D20	90

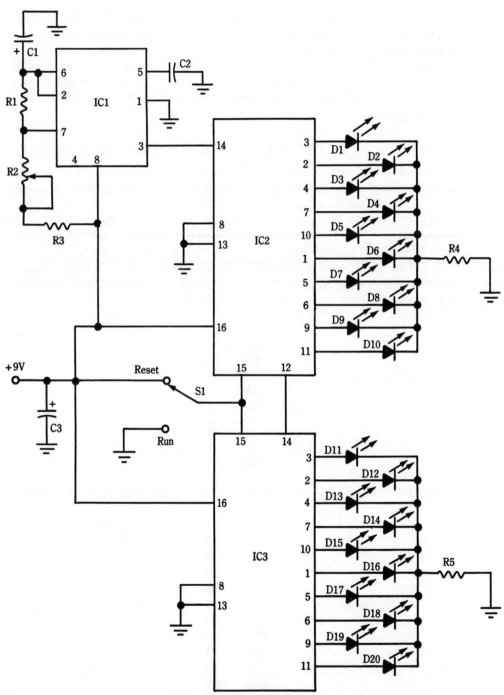

Fig. 13-4. Project 88: extended-range dual-decade counter.

**Table 13-4. Parts List for the Extended-
Range Dual-Decade-Counter Project of Fig. 13-4.**

Component	Description
IC1	7555 (or 555) timer
IC2, IC3	CD4017 decade counter
D1–D20	LED
C1, C3	10-μF, 25-volt electrolytic capacitor
C2	0.01-μF capacitor
R1, R3	10K, 1/4-watt resistor
R2	500K potentiometer
R4, R5	470-ohm, 1/4-watt resistor
S1	SPDT switch

Two LEDs are lit at any given time. Their values simply add together. For example, consider the following combinations:

$$
\begin{array}{lll}
\text{D3 and D15 lit} & = & 42 \\
\text{D1 and D13 lit} & = & 20 \\
\text{D8 and D19 lit} & = & 87 \\
\text{D1 and D20 lit} & = & 00 \\
\text{D6 and D20 lit} & = & 05 \\
\text{D10 and D20 lit} & = & 99
\end{array}
$$

Any value from 00 to 99 can be unambiguously displayed on two of the twenty LEDs in the output display.

As shown in Fig. 13-4, the dual-decade counter goes through its full hundred step range in the count-to-n-and-halt mode. Maximum count values lower than 99 can be set up using the same approach discussed in Project 85. An AND gate is used to trigger the counter when the desired maximum count value is reached.

Refer to Project 86 for an explanation of how the CD4017 can be used in the count-to-n-and-recycle mode.

Potentiometer R2 sets the clock rate. Any external clock can be used in place of IC1 and its associated components (R1, R2, R3, C1, and C2), if you prefer.

PROJECT 89: FLASHING DECIMAL-DIGIT DISPLAY

A flashing LED is much more visible than a continuously lit LED. The same thing holds true for seven-segment displays.

The circuit illustrated in Fig. 13-5 can be used to create a flashing decimal digit display. Table 13-5 shows the parts list. The input to this circuit is a four-line BCD (binary-coded

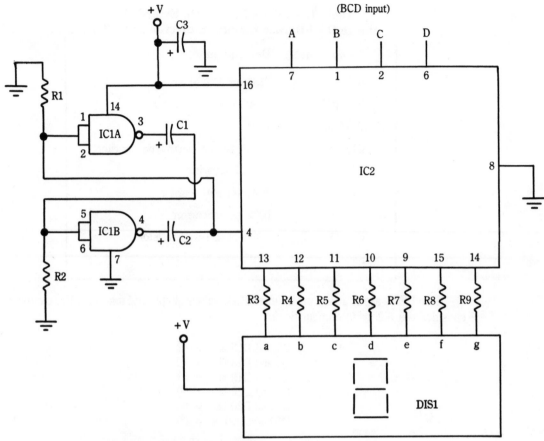

Fig. 13-5. Project 89: flashing decimal-digit display.

**Table 13-5. Parts List for the Flashing
Decimal-Digit-Display Project of Fig. 13-5.**

Component	Description
IC1	CD4011 quad NAND gate
IC2	74C47 BCD-to-7 segment decoder/driver
DIS1	common-anode seven-segment LED display
C1, C2	50-μF, 25-volt electrolytic capacitor
C3	10-μF, 25-volt electrolytic capacitor
R1, R2	3.9K, 1/4-watt resistor
R3–R9	330-ohm, 1/4-watt resistor

decimal) encoded digital value. The output of IC1B can be used to drive several display driver chips (IC2). A buffer stage can be required to increase the fanout.

Experiment with different component values for resistors R1 and R2 and capacitors C1 and C2. The values of these components control the flash rate. Using the component values listed here, the display will flash two or three times per second.

PROJECT 90: MANUALLY PROGRAMMED COUNTDOWN TIMER

Most counter circuits (like all of the projects presented so far in this chapter) count up to some number. That is, the count sequence starts at 0 and increments up on each clock pulse until the maximum count value is achieved. In some applications, a countdown timer is more useful. In a countdown timer, the counting cycle starts with the maximum count value which is then decremented down on each clock pulse until a value of 0 is reached.

A practical countdown timer circuit is illustrated in Fig. 13-6. Table 13-6 shows the parts list for this project.

The counter is incremented only when switch S1 is closed. When this switch is open, the clock pulses (from IC1) cannot reach the counter. The clock rate can be manually adjusted with potentiometer R2. You might also want to experiment with other values for capacitor C1. The larger this capacitance is, the slower the clock rate will be.

The initial (maximum) value is entered via switches S2 through S5 in BCD (binary coded decimal) fashion. A four-unit DIP switch would be a good choice for this project. Acceptable BCD codes that can be entered with these switches are as follows:

BCD	Value
0000	0
0001	1
0010	2
0011	3
0100	4
0101	5
0110	6
0111	7
1000	8
1001	9

The following combinations are possible, but disallowed. They should not be entered, because they will confuse the counting circuit and the output display:

1010
1011
1100
1101
1110
1111

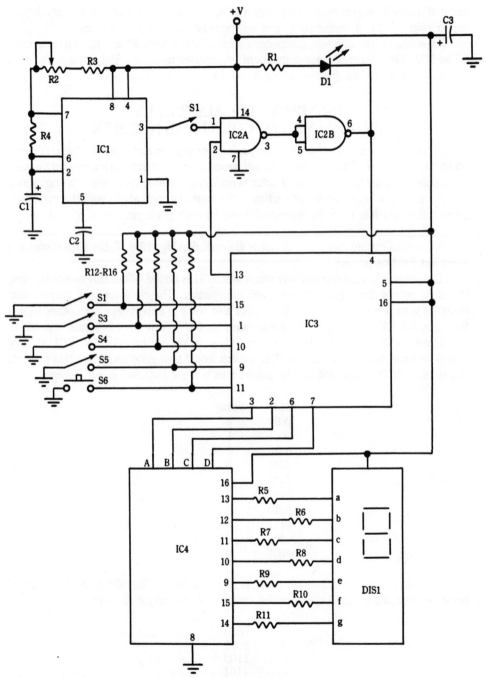

Fig. 13-6. Project 90: manually programmed countdown timer.

**Table 13-6. Parts List for the Manually
Programmed Countdown Timer Project of Fig. 13-6.**

Component	Description
IC1	7555 (or 555) timer
IC2	CD4011 quad NAND gate
IC3	74C192 BCD up/down counter
IC4	74C47 BCD-to-7 segment decoder/driver
D1	LED
DIS1	common-anode seven-segment LED display
C1, C3	10-μF, 25-volt electrolytic capacitor
C2	0.01-μF capacitor
R1	390-ohm, 1/4-watt resistor
R2	500K potentiometer
R3	390K, 1/4-watt resistor
R4	2.2K, 1/4-watt resistor
R5–R11	330-ohm, 1/4-watt resistor
R12–R16	1-megohm, 1/4-watt resistor
S1	SPST switch
S2–S5	SPST switch
S6	normally open SPST pushbutton switch

No permanent damage will be done, however, if you accidentally enter one of these disallowed combinations. Once the counter is reset (via switch S6) and a valid BCD value is entered, everything will work properly. In some cases, you may need to briefly turn the power off and back on to regain control of the circuit.

The output count value is displayed in decimal form on a single-digit, seven-segment display. (DIS1). When a count of 0 is reached, LED D1 also lights up. The counting sequence then stops. To reset the counter for another cycle, just briefly close reset switch S6.

PROJECT 91: RANDOM-NUMBER GENERATOR

By feeding a high-speed clock to a recycling counter through a pushbutton switch, we can create a random-number generator. If the clock rate is fast enough, there is no way to predict what value will be displayed when the switch is released.

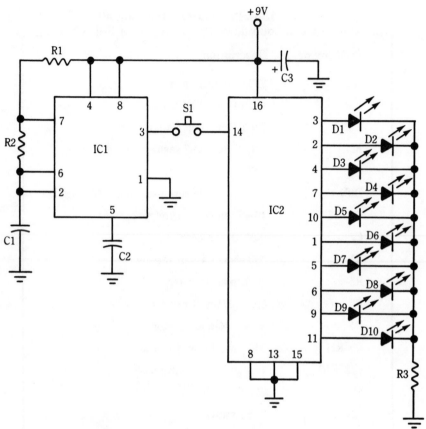

Fig. 13-7. Project 91: random-number generator.

Table 13-7. Parts List for the Random-Number Generator Project of Fig. 13-7.

Component	Description
IC1	7555 (or 555) timer
IC2	CD4017 decade counter
D1–D10	LED
C1, C2	0.01-μF capacitor
C3	10-μF, 25-volt electrolytic capacitor
R1, R2	1K, 1/4-watt resistor
R3	470-ohm, 1/4-watt resistor
S1	normally open SPST pushbutton switch

Figure 13-7 shows a random-number generator circuit using the CD4017 decade counter. The parts list for this project is given in Table 13-7.

The operation of this project is very straightforward. When push switch S1 is closed, the high-speed clock pulses from IC1 increment the counter (IC2) very rapidly. All ten output LEDs will appear to be continuously lit. (The LEDs are actually flashing on and off at a rate too fast to be seen by the human eye.) When the switch is released, only one of the LEDs will remain lit. This is the randomly selected value.

Unlike most counter circuits, an inexpensive pushbutton switch can be used for S1. Any contact bouncing is irrelevant in this particular application. If the switch contacts bounce, that just adds to the randomness of the final output count value.

PROJECT 92: PHOTOELECTRIC COUNTER

So far, all of the counter projects presented in this chapter use optoelectric devices (LEDs) as output devices to indicate the count value. By using a suitable photosensor as the input to a counter, a circuit can actually count objects or events in the real world—not just electrically generated signals.

Figure 13-8 shows the circuit for a photoelectric counter. This is one of the most complex circuits in this book, so work carefully when constructing this project. The parts list is given in Table 13-8.

Depending on how the project is set up mechanically, this circuit can be used as either an object counter or as a rotation counter.

Figure 13-9 illustrates the basic setup for a simple object counter. A light source and the photoelectric sensor are placed across the object path, directly opposite one another. Normally, the light from the source will shine directly on the sensor, but when an object (which can be almost anything) passes between the light source and the sensor, the light beam, as seen by the sensor, is momentarily cut off. The sensor responds to this change in its received light, triggering the counter circuitry. This simple, but versatile system can be used to count objects passing by on a conveyer belt, or people passing through a doorway, to name just two of the many possible applications.

A photoelectric rotation counter works on exactly the same basic principle of blocking the light source from the sensor to trigger the counter. As shown in Fig. 13-10, a light source and the photoelectric sensor are again placed opposite each other in this system. A shaft with an eccentric (not round or evenly proportioned) cam on the end is mounted on the revolving wheel (or whatever revolving object is to be monitored by the counter). The shaft revolves with the wheel. Once per revolution, the wide end of the cam passes between the light source and the sensor, momentarily breaking the light beam. These pulses are counted by the rest of the circuitry. In a sense, this system resembles a mechanical frequency counter.

Referring back to the circuit diagram of Fig. 13-8, transistor Q1 is a phototransistor, serving as the light sensor. IC1 can be almost any op amp, such as the common and inexpensive 741. Potentiometer R4 is adjusted during calibration so that the breaking of the beam of light striking Q1 will cause the comparator (IC1) to emit a clean output pulse.

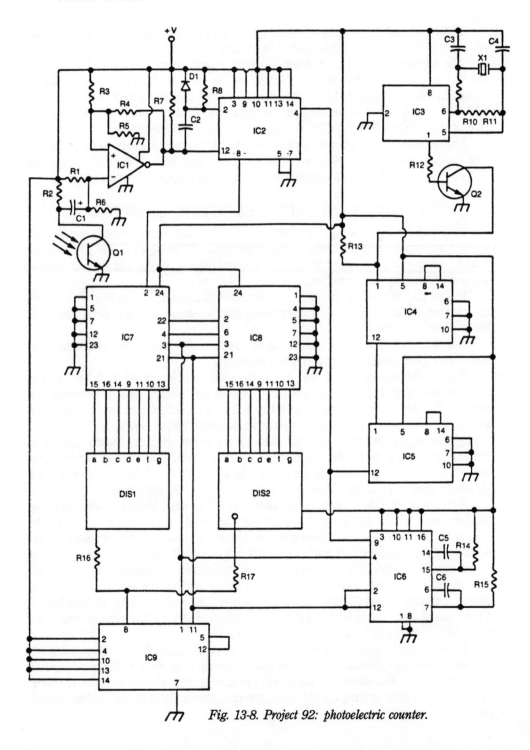

Fig. 13-8. Project 92: photoelectric counter.

Table 13-8. Parts List for the
Photoelectric-Counter Project of Fig. 13-8.

Component	Description
IC1	op amp (741, or similar)
IC2	74C90 J-K flip-flop
IC3	MM5369 60-Hz timebase
IC4, IC5	74C92 divide-by-12 counter
IC6	74C123 dual monostable multivibrator
IC7, IC8	74C143 decade counter/decoder/display driver
IC9	74C74 dual D flip-flop
Q1	phototransistor (FPT-100, or similar)
Q2	npn transistor (2N222, 2N3904, or similar)
D1	diode (1N4148, or similar)
DIS1, DIS2	seven-segment display, common anode (with decimal point)
C1	1-μF, 25-volt electrolytic capacitor
C2	1000-pF capacitor
C3	30-pF capacitor
C4	6.2-pF capacitor
C5, C6	0.033-μF capacitor
R1, R3, R5, R6	220K, 1/4-watt resistor
R2	5.6K, 1/4-watt resistor
R4	2.5-megohm trimpot
R7, R9	1K, 1/4-watt resistor
R8, R14, R15	10K, 1/4-watt resistor
R10, R11	10 megohm, 1/4-watt resistor
R12	15K, 1/4-watt resistor
R13	2.2K, 1/4-watt resistor
R16, R17	330-ohm, 1/4-watt resistor
X1	3.58-MHz (color-burst) crystal

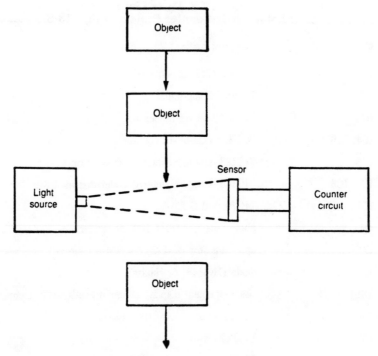

Fig. 13-9. A passing object breaks the light beam, triggering the counter.

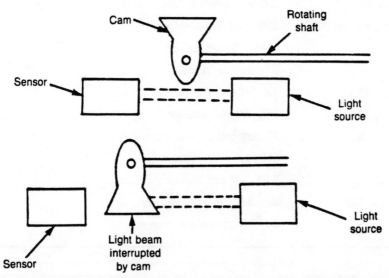

Fig. 13-10. An eccentric cam can be used to convert the photoelectric counter project to a rotation counter.

This pulse is then used to trigger a monostable multivibrator (IC2). This stage of the circuit functions as a "switch" debouncer. Basically, it cleans up any erratic portions of the signal.

IC3 is a timebase oscillator, producing a precise 60-Hz signal. Because the timebase's frequency stability is more important in this application than the actual frequency, the values of the frequency-determining components (capacitors C3 and C4, and resistors R9, R10, and R11) are not terribly important. Use anything that you happen to have on hand that is reasonably close to the values itemized in the parts list.

For some applications (such as counting people passing through a doorway), the timebase oscillator stage can be eliminated. You should also feel free to change the timebase frequency to suit your individual application.

The actual counting procedure is performed by IC4 and IC5. The output display (DIS1 and DIS2) includes a decimal point controlled by IC6 and IC9. Note that the seven-segment LED display units used in this project must be the common-anode type. Common-cathode displays will not work with the 74C143 decode/driver chips used in this project.

Transistor Q2 can be almost any garden-variety npn-type device, such as the 2N3904 or the 2N2222.

While fairly complex in terms of parts count, this circuit is reasonably flexible. I encourage you to experiment with variations on the basic circuit shown here.

Chapter 14
Miscellaneous Projects

Inevitably, when writing a book of projects like this one, a number of useful and interesting circuits are left over. These projects don't fit into any neat chapter categories, but they warrant inclusion in the book.

Rather than having a string of short chapters of one or two projects each, I will wrap up the book with a collection of miscellaneous circuits. I think you'll enjoy them.

PROJECT 93: LIGHT-OFF ALARM

The circuit shown in Fig. 14-1 produces a tone only when the detected light level drops below a specific point. As long as a bright light is shining on the photoresistor (P1), the speaker remains silent. This device can be used to monitor a pilot light, sounding an alarm when the light goes out.

Another possible application would be in a simple intrusion detector. A light beam is directed upon the photoresistor. Anybody passing between the light source and the sensor will temporarily break the light beam, and the alarm will be sounded until the light beam striking the sensor is restored.

A typical parts list for this project is given in Table 14-1. Nothing is very critical in this project. Feel free to experiment with other parts values, especially for the tone determining components (R1, R2, and C1).

Potentiometer R3 is a sensitivity control. It sets the minimum amount of illumination before the tone sounds. Adjust this control for the desired response.

PROJECT 94: LIGHT-ON ALARM

This project is very similar to Project 93. Figure 14-2 shows the schematic, and the parts list appears in Table 14-2.

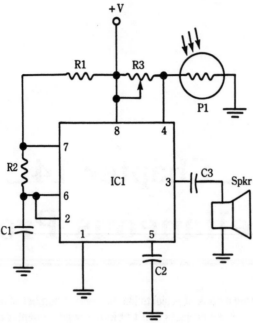

Fig. 14-1. Project 93: light-off alarm.

**Table 14-1. Parts List for the
Light-Off Alarm Project of Fig. 14-1.**

Component	Description
IC1	555 timer
P1	photoresistor (cadmium sulfide)
R1	100K, 1/4-watt resistor
R2	2.2K, 1/4-watt resistor
R3	10K potentiometer
C1, C2	0.01-μF capacitor
C3	0.1-μF capacitor
Spkr	small speaker

The only difference between these two projects is the relative placement of the sensitivity control potentiometer (R3) and the photoresistor (P1). In this project, we have simply reversed their positions. This causes the circuit to operate in the opposite manner from Project 93.

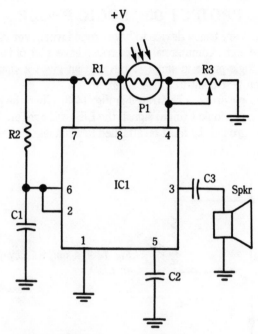

Fig. 14-2. Project 94: light-on alarm.

Table 14-2. Parts List for the Light-On Alarm Project of Fig. 14-2.

Component	Description
IC1	555 timer
P1	photoresistor (cadmium sulfide)
R1	100K, 1/4-watt resistor
R2	2.2K, 1/4-watt resistor
R3	10K potentiometer
C1, C2	0.01-μF capacitor
C3	0.1-μF capacitor
Spkr	small speaker

This project is a light on alarm. The tone sounds when the light striking the sensor is above the threshold level (set by R3). As long as the photoresistor is dark (shielded from external light), the speaker remains silent.

Once again, feel free to experiment with other component values. Nothing in this circuit is particularly critical.

PROJECT 95: LOGIC PROBE

A logic probe is a very handy device to have around whenever you are working with digital circuits. While some commercial logic probes have a lot of fancy features, at the most basic level, a logic probe is simply a device that gives a visual indication of the logic state at a given test point in a digital circuit.

A super-simple logic probe is illustrated in Fig. 14-3. When the probe tip is touched to a connection carrying a logic 1 (high) signal, the LED will light up, otherwise the LED will remain dark. The ground clip must be attached to a suitable ground point in the circuit under test.

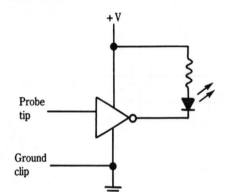

Fig. 14-3. A simple logic probe is little more than an LED.

This circuit could also "steal" its power from the circuit being tested. Just solder an alligator clip on the end of the V+ line, and attach it to the supply voltage of the circuit under test. The current drawn by the logic probe is minimal, and will not cause any loading problems in the vast majority of digital circuits.

While functional, this super-simple circuit might well give ambiguous results under certain circumstances. This is obviously quite undesirable for any piece of test equipment. If the LED is dark, does that mean there is a logic 0 (low) signal, or is the probe not making proper contact with the test point? Does a lit LED indicate a true steady-state logic 1 (high) signal, or a high-speed pulse, causing the LED to blink on and off at a rate too fast for the eye to distinguish the separate flashes? It is also possible that the test point is somehow shorted to the V+ line, causing the LED to light up.

Most of these problems can easily be taken care of by adding a second stage to the logic probe circuit. This is exactly what we will do in this project. Figure 14-4 the circuit, and Table 14-3 gives the parts list.

Essentially, this improved logic probe circuit is just two of the original simple circuits in series. With two LEDs, four possible conditions can be unambiguously indicated:

Both LEDs dark	no input signal
LED1 only lit	logic 1 signal
LED2 only lit	logic 0 signal
Both LEDs lit	high-frequency pulse

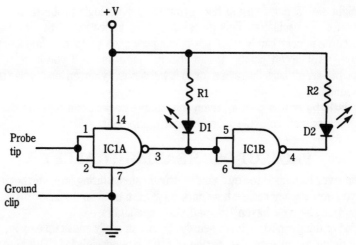

Fig. 14-4. Project 95: logic probe.

**Table 14-3. Parts List for the
Logic-Probe Project of Fig. 14-4.**

Component	Description
IC1	CD4011 quad NAND gate
D1, D2	LED
R1, R2	470-ohm, 1/4-watt resistor

If the signal is a low-frequency pulse, you will be able to see the two LEDs alternately flash on and off.

The only real limitation of this logic probe circuit is that it cannot tell you if the test point is shorted to V+. This condition looks like a steady-state logic 1 signal. A VOM or DMM is necessary to locate this type of problem.

You can add numerous improvements to this simple project. Commercial logic probes often include a number of extra features. You can duplicate many of these in your logic-probe project with just a little bit of effort.

Virtually all of the better commercial logic probes feature pulse stretchers. A *pulse stretcher* is used to indicate more clearly very brief, single pulses. This might sound like a very exotic circuit, but all it really is is a simple monostable multivibrator, or timer. When the pulse stretcher is turned on (it should be switchable, because it will interfere with testing many types of logic signals), it keeps the LED(s) lit for a fixed period of time, even if the input pulse is very short. Very brief input pulses might not visibly light an LED on their own. Such short duration pulses are quite common in many digital circuits and systems.

You might also consider using the output from your logic probe to control a gated clock generator, or oscillator. This gives an audible indication of the logic state. This feature can come in very handy when performing complex tests, or when simultaneously monitoring several circuit points. It can also be quite useful if you have to wait for a single pulse, but aren't sure just when (or if) it will occur. When an input pulse is detected, the tone sounds.

With a little bit of imagination, even this simple project can become the heart of a very sophisticated and versatile piece of test equipment.

PROJECT 96: HEADLIGHT ALERT

Has this ever happened to you? You're driving along during late afternoon/early evening, and you don't really realize how dark its gotten until some other driver blows his horn at you because you haven't turned your headlights on.

This kind of thing can happen very easily. Dusk can really sneak up on you, particularly if you are driving for an extended period of time around nightfall. The human eye will automatically adjust itself for the gradually decreasing light level. You may well not be aware that it is getting dark until something happens to direct your conscious attention to the lighting. You don't want to be alerted to the problem by an accident, or even a near-miss.

The circuit shown in Fig. 14-5 will help you to avoid such problems. Table 14-4 shows the parts list. When the ambient light level drops below a specific level, an LED mounted on your dashboard starts blinking, warning you to turn on your headlights. A blinking LED is hard to miss. When you turn on your headlights, the LED stops blinking and goes dark.

There are two connection points to the car's 12-volt power system. The point marked "+12 volt A" goes to a point after the ignition switch. There is no sense in having the circuit functional when the ignition is off.

The point marked "+12 volt B" is connected to the headlight circuit. A voltage appears on this line only when the headlights are on.

IC1 is a gating network that determines if there is a voltage at point B or not, and if the ambient light level is below the trigger point or not. If the ambient light level is below the trigger point (dark), and there is no voltage at B, then the timer circuit (IC2 and associated components) turns on. The LED (D4) blinks on and off at a rate determined by the values of capacitor C5, and resistors R6 and R7. You might want to experiment with other values for these timing components for the most eye-catching flash rate.

All other conditions (light level above the trigger point and/or a voltage at B), cause the timer and the LED to be cut off.

The sensor for the ambient light level is a simple photoresistor (R4). Potentiometer R5 is a sensitivity control. The setting of R5 determines the trigger level for the ambient light.

For best results, the photoresistor should be remotely positioned (off of the main circuit board) where it can detect the overall ambient lighting level outside the vehicle.

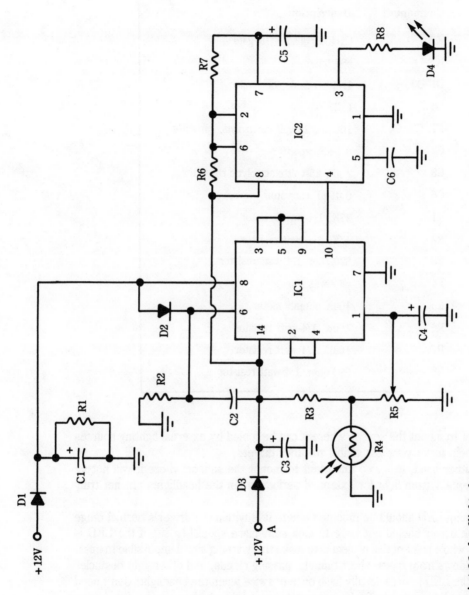

Fig. 14-5. Project 96: headlight alert.

Table 14-4. Parts List for the
Headlight-Alert Project of Fig. 14-5.

Component	Description
IC1	CD4001 quad NOR gate
IC2	555 timer
D1–D3	1N4005 diode, or similar
D4	LED
C1, C3, C4	100-μF, 25-volt electrolytic capacitor
C2	0.1-μF capacitor
C5	2-μF, 25-volt electrolytic capacitor
C6	0.01-μF capacitor
R1	27K, 1/4-watt resistor
R2	47K, 1/4-watt resistor
R3	820-ohm, 1/4-watt resistor
R4	photoresistor
R5	100K potentiometer
R6	150K, 1/4-watt resistor
R7	100K, 1/4-watt resistor
R8	390-ohm, 1/4-watt resistor

Be careful not to mount the sensor where it can be fooled by external lighting sources that do not help make your car visible to other drivers.

On the other hand, it is also important to mount the sensor where it will not be shaded accidentally from light for extended periods when the headlights are not truly needed.

The warning LED should be mounted where it is within the driver's normal range of vision. The driver should not have to look someplace special to see if the LED is blinking. The whole point of this project is to make this warning signal impossible to miss.

Brief shadows from trees, short tunnels, passing clouds, and other light obstacles might cause the LED to occasionally flash once or twice when the headlights don't need to be turned on. However, the flashing stops as soon as the shadow passes. Just ignore such stray, isolated blinks. If the headlights are needed, the LED continues to blink on and off until you turn on the lights.

PROJECT 97: LED OUTPUT TIMER

When most electronics hobbyists need a timer circuit, they almost always reach for a 555 or similar dedicated timer IC. There's certainly nothing wrong with that, but sometimes it is interesting to explore alternate paths to the same end.

Figure 14-6 illustrates a completely different approach to a timer circuit. This timer is built around an adjustable shunt regulator IC, of all things. A typical parts list for this project is given in Table 14-5.

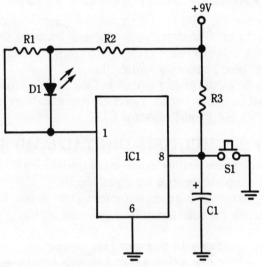

Fig. 14-6. Project 97: LED output timer.

**Table 14-5. Parts List for the
LED Output-Timer Project of Fig. 14-6.**

Component	Description
IC1	TL431 adjustable shunt regulator
D1	LED
C1	22 μF, 25-volt electrolytic capacitor
R1	1.2K, 1/4-watt resistor
R2	330 ohm, 1/4-watt resistor
R3	470K, 1/4-watt resistor
S1	SPST switch (normally open pushbutton)

The time period equation for this timer circuit is a little complex, so it is probably best for most hobbyists to just experiment with different values for timing resistor R3 and timing capacitor C1. Increasing the value of either of these components produces a longer timing period. Of course, the timing period can also be reduced by lowering the value of either R3 or C1, or both.

The timing cycle is initiated by momentarily closing pushbutton switch S1. When the circuit has timed out, the LED lights up.

PROJECT 98: TEN-STEP TIMER

The circuit shown in Fig. 14-7 is a rather unusual timer. It uses a seven-segment display as an output device. The circuit cycles through ten equal timing periods. The seven-segment display tells you which cycle the circuit is currently in. The cycle can be interrupted at any time by opening switch S1.

A complete parts list for this project appears in Table 14-6. Potentiometer R2 permits you to manually adjust the timing period. You might also want to experiment with other values for resistors R1 and R3 and capacitor C1.

PROJECT 99: MULTIBIT DIGITAL COMPARATOR

We have used analog comparators in several of the projects in this book. It is also possible to create a comparator circuit for digital signals.

A simple X-OR gate can act as a comparator for two single bits, but that is of limited value. In this project we will build a multibit digital comparator.

Table 14-6. Parts List for the Ten-Step Timer Project of Fig. 14-7.

Component	Description
IC1	7555 (or 555) timer
IC2	74C90 BCD counter
IC3	74C47 BCD-to-seven-segment decoder/driver
DIS1	seven-segment LED display—common-anode
C1	50-μF, 25-volt electrolytic capacitor
C2	0.01-μF capacitor
C3	10-μF, 25-volt electrolytic capacitor
R1	100K, 1/4-watt resistor
R2	500K potentiometer
R3	2.2K, 1/4-watt resistor
R4–R10	330-ohm, 1/4-watt resistor
S1	SPST switch

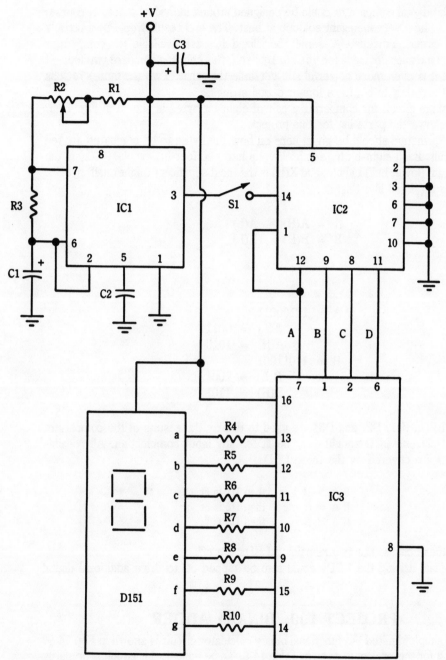

Fig. 14-7. Project 98: ten-step timer.

A multibit digital comparator could be designed around individual gates, of course, but that would be a very inelegant solution at best. The end result would be a terribly complicated circuit. Fortunately, some specialized ICs are available for comparator applications. One such device is the 74C85. By itself, this chip can compare two four-bit nybbles, but it is even more powerful and versatile than that. Two (or more) 74C85s can be used together to compare longer digital strings.

A two-stage circuit for comparing a pair of eight-bit bytes is shown in Fig. 14-8. Table 14-7 shows the parts list for this project.

Some explanation should be given here on how the bytes to be compared are fed into the circuit. Each eight-bit byte is broken up into two four-bit nybbles. X(L) is the least significant (lower half) nybble, and X(H) is the most significant (higher half) nybble. Each byte is arranged like this:

$$A = A(H) \quad A(L)$$
$$B = B(H) \quad B(L)$$

For example,

$$A = 10100011$$
$$A(L) = 0011$$
$$A(H) = 1010$$
$$B = 11011001$$
$$B(L) = 1001$$
$$B(H) = 1101$$

Three LEDs (D1, D2, and D3) are used to display the results of the comparator operation. Only one LED should ever be lit for any input combination. All possible combinations are covered by the three LEDs:

If A > B then D1 is lit
If A = B then D2 is lit
If A < B then D3 is lit

In each of these cases, the two remaining LEDs are off.

The signals driving the LEDs could also be tapped off to drive additional digital circuitry.

PROJECT 100: BINARY ADDER

A very simple, limited but functional binary calculator circuit is shown in Fig. 14-9. The parts list for this project is given in Table 14-8. To be honest, this circuit is primarily for demonstration and educational purposes. It isn't really all that useful in and of itself, but what you can learn from this project could be used in more advanced projects of your own design.

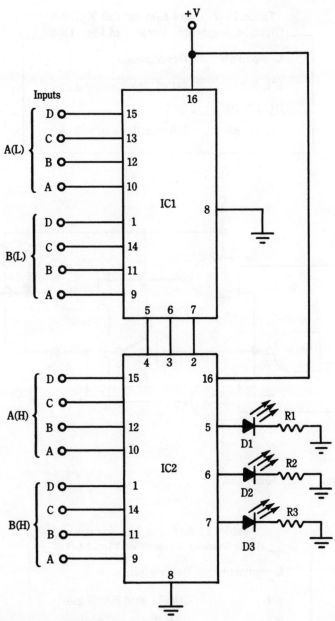

Fig. 14-8. Project 99: multibit digital comparator.

This circuit uses all four gates in a CD4011 quad NAND gate IC. IC2 is listed as a CD4049 hex inverter in the parts list, but if you prefer, you could substitute one section of a second CD4011 quad NAND gate (or a CD4001 quad NOR gate). Just short the inputs together and use the gate as an inverter. It really boils down to which chip you

Table 14-7. Parts List for the Multibit Digital-Comparator Project of Fig. 14-8.

Component	Description
IC1, IC2	74C85 digital comparator
D1, D2, D3	LED
R1, R2, R3	330-ohm, 1/4-watt resistor

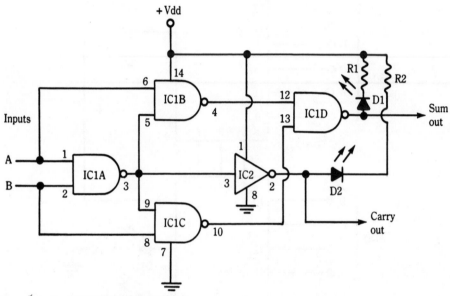

Fig. 14-9. Project 100: binary adder.

Table 14-8. Parts List for the Binary-Adder Project of Fig. 14-9.

Component	Description
IC1	CD4011 quad NAND gate
IC2	CD4049 hex inverter
D1, D2	LED
R1, R2	330-ohm, 1/4-watt resistor

happen to have handy. Such a substitution will not affect the performance of the circuit in any way.

The binary adder project adds two single digit (one-bit) binary numbers. With two one-bit inputs there are four possible input combinations:

$$0 + 0$$
$$0 + 1$$
$$1 + 0$$
$$1 + 1$$

The results for the first three combinations should be pretty obvious, because they are the same as in the common decimal number system we are all familiar with:

$$0 + 0 = 0$$
$$0 + 1 = 1$$
$$1 + 0 = 1$$

The fourth combination (1 + 1) might throw you if you are not familiar with the binary numbering system. In decimal, $1 + 1 = 2$, but there is no digit "2" in the binary number system. To express values larger than one, we must start a new column, just as in the decimal system a second column is necessary to write values higher than nine.

Therefore, a functional binary-added circuit needs a secondary output for the carry digit to create (or be added to) the next higher column. In binary,

$$1 + 1 = 10$$

The two LEDs (D1 and D2) display the output states. A lit LED represents a 1, and a dark LED represents a 0. These are the only possible digits in the binary number system.

In truth table form, the operation of this circuit can be summarized as follows:

INPUTS		OUTPUTS	
A	B	Sum	Carry
0	0	0	0
0	1	1	0
1	0	1	0
1	1	0	1

Technically speaking, the circuit we are using in this project is called a *binary half-adder*. This terminology is used because the circuit does not have a carry input from a previous column. As an educational exercise you might want to try designing a two-digit binary full-adder. Remember, the circuitry for the second column must handle three inputs:

1. A's most significant (second column) bit
2. B's most significant (second column) bit
3. The carry from adding the least significant (first column) bits

Believe it or not, this simple circuit (or something very similar) is at the heart of even the most sophisticated digital computer in the world.

PROJECT 101: LIGHT-ACTIVATED DIGITAL GATE

Many of the projects in this book use a photosensor to feed information from the outside world into an analog circuit. Photosensitive devices can also be used to supply digital signals.

Figure 14-10 shows a simple light-activated digital-gate circuit. A parts list for this project is given in Table 14-9.

The sensor in this circuit is a simple photoresistor. Potentiometer R1 is a sensitivity control. It is used to adjust the trip point of the circuit. That is, R1 controls how much light is required to trigger the circuit.

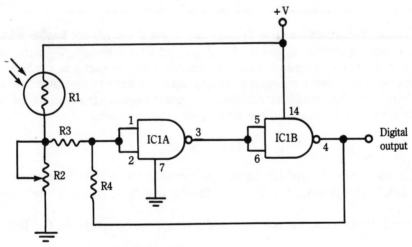

Fig. 14-10. Project 101: light-activated digital gate.

Table 14-9. Parts List for the Light Activated Digital-Gate Project of Fig. 14-10.

Component	Description
IC1	CD4011 quad NAND gate
R1	photoresistor
R2	100K potentiometer
R3	10K, 1/4-watt resistor
R4	120K, 1/4-watt resistor

The output goes high when photoresistor R1 is illuminated. If R1 is dark (in shadow), the output will go low. It is easy enough to adapt the circuit for the opposite response. Just add an extra inverter stage at the output.

PROJECT 102: AUTOMATIC NIGHT LIGHT (BONUS PROJECT)

Have you ever came home late at night when you've forgotten to turn on the porch light? Isn't it frustrating trying to fit the key into the lock in the dark?

Or have you ever gotten up in the middle of the night to go to the bathroom and tripped over one of the kids' toys, or the cat? (My cat is black, making this kind of occurrence even more likely.) Then you have to fumble around for the light switch.

A small night light can make life easier, but it is ridiculous (and wasteful) to leave it burning during daylight hours. But who can remember to turn the darn thing on every evening and off every morning? Clearly, this is an ideal job for automation.

You could rig up a timer of some sort, but in this case that would be overkill. The exact time isn't at all important. When it's dark, we want the night light to come on. When there is sufficient light, the night light should be turned off.

Of course, a photoelectric sensor is exactly what we need. A circuit for an automated night light appears in Fig. 14-11. Table 14-10 shows the parts list.

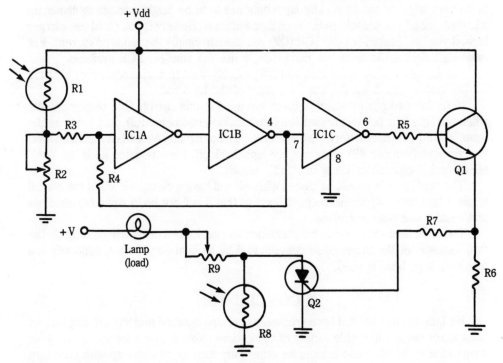

Fig. 14-11. Project 102: automatic night light (bonus project).

**Table 14-10. Parts List for the
Automatic Night-Light Project of Fig. 14-11.**

Component	Description
IC1	CD4049 hex inverter
Q1	npn transistor (2N3904, 2N2222, or similar)
Q2	SCR—select to suit load (lamp)
R1, R8	photoresistor
R2, R9	100K potentiometer
R3	10K, 1/4-watt resistor
R4	100K, 1/4-watt resistor
R5, R6	1K, 1/4-watt resistor
R7	120-ohm, 1/4-watt resistor

The lamp is simply a small dc bulb, like those used in flashlights. It doesn't have to be very large or powerful. The lamp only needs to be bright enough to illuminate a keyhole, or let you see where the main light switch is. However, you could use a larger bulb, if you like. Make sure the SCR (Q2) you use can handle the required current. For most night-light applications you can probably use the smallest SCR available.

For use as an indoor night light you will probably want to devise some sort of shade for the lamp to cut down the annoying glare of the bare bulb.

Note that two photoresistor sensors are used in this circuit. One sensor detects when the light falls below a preset level (set via potentiometer R2), and turns on the lamp. The other sensor determines when the ambient lighting exceeds a minimum level (set via potentiometer R9), and turns the lamp back off. Photoresistor R1 is the "on" sensor and photoresistor R8 is the "off" sensor.

The "on" sensor should be placed where it will have a clear "view" of the overall ambient light level. R1 should be positioned so that it will not be in any stray shadows that could cause false triggering.

The night light itself should be positioned so that it does not shine directly on the "off" sensor, or the circuit could get confused by its own output light, especially if a relatively large lamp is used.

IN CONCLUSION

Well, there you have it. I have presented over one hundred projects utilizing various optoelectric devices in a wide variety of applications. Now it's time for you to use your imagination to modify these circuits for other purposes, or to come up with your own exciting optoelectric projects.

Index

A

ac-powered flasher, 109
adder, binary, 218-222
alternate light Morse code receiver, 157
alternate light-range detector, 175
alternate peak detector, 135-136
alternate simple light meter, 175
amplifier, light-controlled, 68-70
audio wattmeter, 131
automatic light-balance controller, 66-67
automatic night light, 223-224

B

bargraph display, 20-22, 117-118
  blinking over-range alert, 216
  light meter with, 177-179
  over-range alert and, 122-126
  ten-step, 118-121
  variable-range, 121
basic LED flasher, 93-94
batteries, 4
  solar, 5
battery pack
  high-current solar, 44-45
  solar, 41-44
bidirectional diodes, 34
binary adder, 218-222
binary counter demonstrator, 185-188
binary half-adder, 221
bird-chirp simulator, light-controlled, 80-81
breakdown, 14

C

capacitance, light-controlled, 59
chaser, light, 114-116
chopped light-activated relay, 58-59
Christmas tree, electronic, 102-106
coaxial cable, fiber optics vs., 38
coin flipper, electronic, 149-152
common-cathode displays, 19
communications projects, 155-170
  FM optical receiver, 168
  light Morse code receiver, 156
  light Morse code transmitter, 155
  optical FM transmitter, 162-163
  optical receivers, 163-169
  optical voice transmitter, 160-161
  simple tone transmitter, 158-159
  two-tone transmitter, 159-160
comparator
  multibit digital, 218
  phase and frequency, 127-130
components, 1-38
control circuits, 53-70
  automatic light-balance controller, 66-67
  chopped light-activated relay, 58-59
  dark-controlled relay, 54-56
  heavy duty light-activated relay, 56-58
  light beam interruption detector, 62-63
  light dimmer, 65
  light-controlled amplifier, 68-70
  light-controlled capacitance, 59-61
  light-controlled motor, 61-62
  light-operated relay, 53-54
  self-activating night light, 63-65
  universal voltage control, 67
countdown timer, manually programmed, 197-199
counter, photoelectric, 201-205
counter circuits, 185-205
  binary counter demonstrator, 185-188
  extended-range dual-decade counter, 193-195
  flashing decimal-digit display, 195-197
  manually programmed countdown timer, 197-199
  photoelectric counter, 201-205
  random-number generator, 199-201
  repeating-cycle variable-length counter, 191-193
  single-cycle variable-length counter, 188-190

D

dark-activate flasher, 110-111
dark-controlled relay, 54-56
decimal-digit display, flashing, 195-197
deluxe light-controlled rectangle-wave generator, 84-86
detector, light beam interruption, 62-63
dice, electronic, 137-140
digital solar theremin, 90-92
dimmer, 65
diodes, photocells and, 5
dot-graph display, ten-step, 121
dry cells, 4

Other Bestsellers From TAB

☐ **SUPERCONDUCTIVITY—THE THRESHOLD OF A NEW TECHNOLOGY—Jonathan L. Mayo**

Superconductivity is generating an excitement in the scientific world not seen for decades! Experts are predicting advances in state-of-the-art technology that will make most existing electrical and electronic technologies obsolete! This book is one of the most complete and thorough introductions to a multifaceted phenomenon that covers the full spectrum of superconductivity and superconductive technology. 160 pp., 58 illus.

Paper $14.95   **Hard $18.95**
Book No. 3022

☐ **FIBEROPTICS AND LASER HANDBOOK—2nd Ed.—Edward L. Safford, Jr. and John A. McCann**

Explore the dramatic impact that lasers and fiberoptics have on our daily lives—PLUS, exciting ideas for your own experiments! Now, with the help of experts Safford and McCann, you'll discover the most current concepts, practices, and applications of fiberoptics, lasers, and electromagnetic radiation technology. Included are terms and definitions, discussions of the types and operations of current systems, and amazingly simple experiments you can conduct! 240 pp., 108 illus.

Paper $19.95   **Hard $24.95**
Book No. 2981

☐ **BUILD YOUR OWN LASER, PHASER, ION RAY GUN AND OTHER WORKING SPACE-AGE PROJECTS—Robert E. Iannini**

Here's the highly skilled do-it-yourself guidance that makes it possible for you to build such interesting and useful projects as a burning laser, a high power ruby/YAG, a high-frequency translator, a light beam communications system, a snooper phone listening device, and much more—24 exciting projects in all! 400 pp., 302 illus.

Paper $13.95   **Book No. 1604**

☐ **BUILD YOUR OWN WORKING FIBEROPTIC, INFRARED AND LASER SPACE-AGE PROJECTS—Robert E. Iannini**

Here are plans for a variety of useful electronic and scientific devices, including a high sensitivity laser light detector and a high voltage laboratory generator (useful in all sorts of laser, plasma ion, and particle applications as well as for lighting displays and special effects). And that's just the beginning of the exciting space age technology that you'll be able to put to work! 288 pp., 198 illus.

Paper $18.95   **Hard $24.95**
Book No. 2724

☐ **20 INNOVATIVE ELECTRONICS PROJECTS FOR YOUR HOME—Joseph O'Connell**

O'Connell carefully guides the budding inventory and enhances the ability of the experienced designer. This book is a no-nonsense approach to building unusual yet practical electronic devices. More than just a collection of 20 projects, this book provides helpful hints and sound advice for the experimenter and home hobbyist. Particular emphasis is placed on unique yet truly useful devices that are justifiably time- and cost-efficient. Projects include a protected outlet box (for your computer system) . . . a variable ac power controller . . . a remote volume control . . . a fluorescent bike light . . . and a pair of active minispeakers with built-in amplifiers. 256 pp., 130 illus.

Paper $15.95   **Hard $21.95**
Book No. 2947

☐ **THE ROBOT BUILDER'S BONANZA: 99 INEXPENSIVE ROBOTICS PROJECTS—Gordon McComb and John Cook**

Where others might only see useless surplus parts you can imagine a new "life form." Now, there's a book that will help you make your ideas and dreams a reality. With the help of the *Robot Builder's Bonanza* you can truly express your creativity. This fascinating guide offers you a complete, unique collection of tested and proven project modules that you can mix and match to create an almost endless variety of highly intelligent and workable robot creatures. 336 pp., 283 illus.

Paper $17.95   **Hard $23.95**
Book No. 2800

☐ **101 SOLDERLESS BREADBOARDING PROJECTS—Delton T. Horn**

Would you like to build your own electronic circuits but can't find projects that allow for creative experimentation? Want to do more than just duplicate someone else's ideas? In anticipation of your needs, Delton T. Horn has put together the ideal project *ideas* book! It gives you the option of customizing each project. With over 100 circuits and circuit variations, you can design and build practical, useful devices from scratch! 220 pp., 273 illus.

Paper $18.95   **Hard $24.95**
Book No. 2985

☐ **BEYOND THE TRANSISTOR: 133 ELECTRONICS PROJECTS—Rufus P. Turner and Brinton L. Rutherford**

Strongly emphasized in this 2nd edition are the essential basics of electronics theory and practice. This is a guide that will give its reader the unique advantage of being able to keep up to date with the many rapid advances continuously taking place in the electronics field. It is an excellent reference for the beginner, student, or hobbyist. 240 pp., 173 illus.

Paper $12.95   **Hard $16.95**
Book No. 2887

Other Bestsellers From TAB